Scale and Geographic Inquiry

Dedication

This book is dedicated to the faculty, staff and students who have participated in, and contributed to the success of, the Geography Department at the University of Minnesota during its first 77 years, and to our daughters Katherine, Keiko, and Kirstin.

Scale and Geographic Inquiry

Nature, Society, and Method

Edited by

Eric Sheppard

and

Robert B. McMaster

350 Main Street, Malden, MA 02148-5020, USA
108 Cowley Road, Oxford OX4 1JF, UK
550 Swanston Street, Carlton, Victoria 3053, Australia

First published 2004 by Blackwell Publishing Ltd

Library of Congress Cataloging-in-Publication Data

Scale and geographic inquiry : nature, society, and method / edited by
Eric Sheppard and Robert B. McMaster.
p. cm.
Includes bibliographical references and index.
ISBN 0-631-23069-6 (alk. paper) – ISBN 0-631-23070-X (pbk. : alk. paper)
1. Geography–Mathematics. 2. Multidimensional scaling. 3. Geographical perception.
I. Sheppard, Eric S. II. McMaster, Robert Brainerd.

G70.23.S32 2003
910′.01–dc21

2003004176

A catalogue record for this title is available from the British Library.

Set in 10/12.5pt Plantin
by Kolam Information Services Pvt. Ltd, Pondicherry, India

For further information on
Blackwell Publishing, visit our website:
http://www.blackwellpublishing.com

Contents

Figures

Tables

Contributors

Dwight Brown is Professor of Geography at the University of Minnesota where he has authored and coauthored (with Philip Gersmehl and Kyle Brown) papers on biogeography and grassland restoration in the *Annals of the Association of American Geographers*, and a textbook *Alternative Biogeographies of the Global Garden* (with Philip Gersmehl and Susy Ziegler) (in press).

Thomas W. Crawford is an Assistant Professor in the Department of Geography at East Carolina University. His research examines interconnections between geographic information science, population–environment interactions, land-use/land-cover characterization and analysis, and complex systems science.

Kelley A. Crews-Meyer is Assistant Professor and Director of Geographic Computing Systems in the Department of Geography at the University of Texas. She is founder and Director of UT's Geographic Information Science Center, and Associate Director of UT's Environmental Science Institute. Her research interests lie at the intersection of remote sensing of environment, GIScience, quantitative landscape ecology, and environmental policy. Recent publications include Linking People, Place, and Policy: A GIScience Approach (2002, coedited with Stephen J. Walsh) and articles in *Geocarto International; Agriculture, Ecosystems, and Environment*; and *Photogrammetric Engineering and Remote Sensing*.

William E. Easterling is Professor of Geography and Founding Director of the Institutes of the Environment at The Pennsylvania State University. He edited *Making Climate Forecasts Matter* (with Paul Stern, 1999) published by the National Research Council/National Academy Press and published numerous scholarly articles in the past five years on scale and environmental change in journals such as *Climatic Change, Global Environ-*

mental Change, Ecological Modeling, Agronomy Journal, and *Agricultural and Forest Meteorology.*

Michael F. Goodchild is Professor of Geography at the University of California, Santa Barbara; Chair of the Executive Committee of the National Center for Geographic Information and Analysis; Director of the Center for Spatially Integrated Social Science; and Associate Director of the Alexandria Digital Library. His research focuses on geographic information science, and he is author or editor of close to 350 papers and ten books, including, most recently, *Uncertainty in Geographical Information* (2002) with Jingxiong Zhang, and *Fundamentals of Geographic Information Science* (2003) with Matt Duckham and Michael Worboys.

Nina Siu-Ngan Lam is Richard J. Russell Louisiana Studies Professor of Geography at Louisiana State University, where she has coedited *Fractals in Geography* (with Lee De Cola, 1993 and reprinted in 2002). She has published in a number of journals on related research areas including spatial and areal interpolation, spread of the AIDS epidemic, uncertainties in human health risk assessment, and spatial analysis for land-cover change detection.

Helga Leitner is Professor of Geography and a faculty member in the Institute for Global Studies and the Interdisciplinary Center for the Study of Global change at the University of Minnesota. She has published two books, and has written numerous articles and book chapters on the political economy of urban development and urban entrepreneurialism, the politics of citizenship and immigrant incorporation, and the politics of scale. Her current research interests include geographies of governance and citizenship, and interurban policy and activist networks.

Sallie Marston is Professor of Geography and Regional Development at the University of Arizona. She is coeditor of *Life's Work* (with Katharyne Mitchell and Cindi Katz, 2003), *Making Worlds* (with Susan Aiken, Ann Brigham, and Penny Waterstone, 1998), and coauthor of *Places and Regions in Global Context* (with Paul Knox, 2004) and *World Regions in Global Context* (with Paul Knox and Diana Liverman, 2002).

Robert B. McMaster is Professor of Geography and Associate Dean for Planning in the College of Liberal Arts at the University of Minnesota. His area of research include multiple scale databases and cartographic generalization; GIS and society, including environmental risk assessment and public participation GIS (PPGIS); and the history of US academic cartography. He has published two books on map generalization, and was an associate editor on the recently published *Manual of Geospatial Science and*

Technology. From 1990 to 1996, he served as editor of *Cartography and Geographic Information Science*, and is currently a Vice President of the International Cartographic Association.

Gary Pereira is Assistant Professor of Geography at San José State University. He specializes in remote sensing and physical geography. He has published in Geographic Analysis.

Jonathan D. Phillips is Research Challenge Trust Fund Professor of Geography at the University of Kentucky. He has published extensively on the results of field-based research in geomorphology, pedology, and hydrology, mainly in various tick-infested areas of the American South.

Colin Polsky is a geographer specializing in the human dimensions of global environmental change, emphasizing the statistical analysis of vulnerability to climate change, and the blending of quantitative and qualitative methods. Dr. Polsky is finishing a two-year Postdoctoral Fellowship at Harvard University, with the Research and Assessment Systems for Sustainability program at the Belfer Center for Science and International Affairs, John F. Kennedy School of Government. He joins the faculty of the Graduate School of Geography at Clark University (Worcester, Massachusetts) in 2003.

Neil Smith is Distinguished Professor of Anthropology and Geography, at the Graduate Center of the City University of New York where he also directs The Center for Place, Culture and Politics. He works on the broad connections between space, social theory, and history, and his books include *American Empire: Roosevelt's Geographer and the Prelude to Globalization* (2003), *New Urban Frontier: Gentrification and the Revanchist City* (1996) and *Uneven Development: Nature, Capital and the Production of Space* (1991). Author of more than 120 articles and book chapters, his work has been translated into ten languages. He has been awarded Honors for Distinguished Scholarship by the Association of American Geographers and a John Simon Guggenheim Fellowship, and is an organizer of the International Critical Geography group.

Eric Sheppard is Fesler-Lampert Professor in Geography, and adjunct to the Interdisciplinary Center for Global Change, the Center for Advanced Feminist Studies and the Department of American Studies, at the University of Minnesota. He has coauthored *The Capitalist Space Economy* (with T. J. Barnes, 1990), *A World of Difference* (with P. W. Porter, 1998), coedited *A Companion to Economic Geography* (with T. J. Barnes, 2000), and published over 80 refereed articles and book chapters. His current research interests include spatiality and political economy, environmental justice, critical GIS, and interurban policy and activist networks.

Erik Swyngedouw is University Reader in Economic Geography and Fellow of St. Peter's College, Oxford University. He has worked extensively on questions of spatial political economy and on political ecology. Recent books include *The Urbanisation of Injustice* (coedited with Andy Merrifield, 1997), *Urbanising Globalisation* (with Frank Moulaert and Arantxa Rodriguez, 2003), *Flows of Power: The Political Ecology of Water in Guayaquil, Ecuador* (2003). His forthcoming book is *Glocalisations and the Politics of Scale* (2004).

Peter J. Taylor is Professor of Geography, and Co-Director of the Globalization and World Cities (GaWC) study group and network, at Loughborough University. He was founding editor of *Political Geography* and *Review of International Political Economy*. Recent books include *World City Network: A Global Urban Analysis* (forthcoming); *Geographies of Global Change: Remapping the World* (coedited with R. J. Johnston and M. Watts, 2003), and *Political Geography: World-Economy, Nation-State and Locality* (4th edn with Colin Flint, 1999).

Stephen J. Walsh is Professor of Geography, member of the Ecology Curriculum, and Research Fellow of the Carolina Population Center at the University of North Carolina – Chapel Hill, formerly Amos H. Hawley Professor of Geography and Director of the Spatial Analysis Unit at the Carolina Population Center. He has coedited: *GIS and Remote Sensing Applications in Biogeography and Ecology* (2001); *Linking People, Place, and Policy: A GIScience Approach* (2002); and *People and the Environment: Approaches for Linking Household and Community Surveys to Remote Sensing and GIS* (2003). Specific research foci are on pattern and process at the alpine treeline ecotone, biocomplexity, scale dependence and information scaling, land-use and land-cover dynamics, and population–environment interactions.

William F. Welsh is Assistant Professor of Geography at the University of North Carolina at Greensboro. His research specialties include geographic information science, remote sensing, and human–environment relationships. His current research focuses on land-use/land-cover change and land degradation in Nang Rong, Thailand, and urban/suburban sprawl patterns and associated environmental impacts in North Carolina.

Susy S. Ziegler is Assistant Professor of Geography at the University of Minnesota. She has published articles on forest dynamics and disturbance in *Canadian Journal of Forest Research, Global Ecology and Biogeography*, and *American Midland Naturalist*.

Preface

The papers collected in this volume were originally commissioned as a series of public lectures celebrating the 75th anniversary of the Geography Department at the University of Minnesota during the spring of 2000. The Geography Department at the University of Minnesota is the fifth oldest in the United States, founded in 1925 by Darrell Haug Davis who had recently moved from the University of Michigan. Richard Hartshorne joined in 1924, followed by Ralph Hall Brown and Samuel Dicken, who together established Minnesota's reputation as a center for scholarship in historical, philosophical, and human geography. Hartshorne's *Nature of Geography* (1938), and Brown's *Mirror for Americans* (1941) and *Historical Geography of the United States* (1948) became classics in their fields. Major changes occurred after World War II with the retirement of Davis, the departure of Hartshorne and Dicken, and the death of Brown. Renewal of the department occurred under Jan Broek, and the intellectual leadership of John Borchert and Fred Lukermann, during which time the department expanded its scholarly profile to incorporate physical geography. John Fraser Hart, E. Cotton Mather, Philip W. Porter, Joe Schwartzberg, and Yi Fu Tuan established the department's national reputation as a center for cultural geography while Richard Skaggs and Dwight Brown established the department's biophysical program. Later, during the 1960s and 1970s the department, partially because of its situation in a thriving metropolitan region, developed considerable depth in urban geography with John Borchert and John Adams and other faculty members at the University of Minnesota. The department, now with some 22 faculty members, provides undergraduate and graduate instruction that emphasizes a broad education in human, physical, environmental geography, and geographic information science/systems, stressing both a strong theoretical and a rigorous quantitative and qualitative empirical training in the discipline. Current areas of strength include urban and economic geography, cultural geography, nature–society relations, geographic information science, GIS and society, climate and biogeography, and geographic education.

The topic of the lecture series, "Scale and Geographic Inquiry," was chosen to reflect the department's reputation as a broad-based community of geographers with an abiding interest in the nature of geographic inquiry. Geographic scale has received considerable scholarly attention across the discipline in recent years, making it an ideal focus for examining the range of geographic inquiry. We invited as speakers a mix of geographers representing the breadth of the field, each a leading researcher on questions of geographic scale over the last decade. Each author gave a lecture and an informal seminar with faculty and students, and was asked to provide a chapter for this volume. We also commissioned a chapter on scale in biogeography to balance other contributions in physical geography.

The result is a set of essays by leading researchers that demonstrate the depth and breadth of scholarship on geographic scale, which we hope provides a definitive assessment of the field and a benchmark for further work on geographic scale in and beyond geography. While we began with the idea of categorizing these essays as either human, biophysical, or methods, in fact many defy such categorization. For example, Walsh et al. (Chapter 2) embrace methods and biophysical geography, Goodchild (Chapter 7) discusses cartography and human geography, and Swyngedouw (Chapter 6) applies a human geographic approach to environmental geography. One of the distinguishing features of geography over the last 40 years, emphasized at Minnesota, has been its ability to embrace an exceptionally broad range of epistemologies, methodologies, and topics, eschewing a canonical approach to the discipline. As the chapters that follow demonstrate, this diversity can create tensions between what may appear to be fundamentally different approaches to geographic scale. Yet, as we seek to show in our introductory and concluding essays, the diversity hides considerable overlap. Geography's vitality depends on mutual respect and cross-fertilization between its different proponents, and certainly our understanding of geographic scale can only be enriched by engagement across, and not just within, the different approaches to the topic collected in this volume.

Introduction: Scale and Geographic Inquiry

Robert B. McMaster and Eric Sheppard

The concept of geographic scale has intrigued scholars from many disciplines for centuries. From science to fiction, authors have struggled with the many meanings and problems in understanding geographic scale. In his novel *Sylvie and Bruno Concluded*, Lewis Carroll (1893) provides a lay example of the importance of scale:

> "That's another thing we've learned from your Nation," said Mein Herr, "map-making. But we've carried it much further than you. What do you consider the largest map that would be really useful?"
> "About six inches to the mile."
> "Only about six inches!" exclaimed Mein Herr. "We very soon got to six yards to the mile. Then we tried a hundred yards to the mile. And then came the grandest idea of all! We actually made a map of the country on the scale of a mile to the mile!"
> "Have you used it much?" I enquired.
> "It has never been spread out yet," said Mein Herr: "the farmers objected: they said it would cover the whole country and shut out the sunlight! So we now use the country itself, as its own map, and I assure you it does nearly as well." (Carroll, 1893)

Scale is intrinsic to nearly all geographical inquiry. It has received increasing attention within geography in recent years, with significant differences in the understanding of scale emerging among the subdisciplines. Geography's cognate disciplines, including ecology, meteorology and climatology, geology, economics, sociology, and political science, also have strong interests in the concept of spatial scale. Indeed it is difficult to identify a completely "scaleless" discipline. Whereas quantum physicists deal with scaled relations between quarks, neutrons, and atoms, and medicine and the emerging work in genomics is involved with mapping and scale at the level of genes, astronomers operate at the other extreme, conceptualizing space both in terms of light-year distances and alternative geometries.

Traditionally, geographers thought about scale as predominantly a cartographic concept, where scale associates a map distance of a feature to that feature's distance on the surface of the earth. This representative fraction (RF) has become the standard method for representing this meaning of scale. As discussed below, this definition of scale is mathematical, and remains the focus of the cartographer. Robinson and Petchenik note restrictions on this focus:

> Cartographers are not concerned with mapping at all scales of spatial relation. The arrangement of the components of a molecule of DNA, for example, may obviously be "mapped"; this molecule occupies space on the earth, and such a mapping activity might seem to have a logical counterpart of the mapping of the arrangement of streets within a city. In common usage, however, such sub-microscopic mapping lies outside the activity of the cartographer, as do the scales of architectural and engineering drawing. (Robinson and Petchenik, 1976: 53)

Biophysical geographers rely heavily on mathematics, but are concerned with the ranges of "operational" scales in which processes operate, and often consider scales as nested. A classic example is a river's drainage basin, which can be subdivided into the smaller scale watershed of its tributaries, each of which can be divided again into even smaller scale watersheds of the tributaries' tributaries. "Contemporary human geographers are drawn increasingly to diverse scales of study because of the wide range of subjects they address, and also because of their use of explanatory modes in which sensitivity to scale effects is explicit, modes that themselves imply spatial meaning" (Meyer et al., 1992: 257). If biophysical geographers are concerned with the scale dependence of phenomena and processes, and with finding the principles and laws that operate at different scales, human geographers increasingly view scale more as a social construction than a concept guided by definitive laws. In this view, scale is not an externally given attribute of human processes, nor do particular processes necessarily operate at characteristic scales, making the mathematical modeling of scale problematic. Scales are thus spatially and temporally fluid. Nation states change their scale, such as when the Soviet Union subdivided or Germany reunited. Furthermore, whereas social scientists used to think of nation-states as the predominant scale at which political process govern society, it is now commonly argued that globalization has made supra and subnational scales just as important operational scales for governance.

In short, different concepts of scale are employed in geography's various subdisciplines, making any modern definition difficult. Although much has been written recently on scale in geography, there has been little attempt to integrate across these subdisciplinary perspectives. The purpose of this edited volume is to address this failure, by comparing and contrasting the different approaches within geography. In this introduction we seek to provide a context for the essays that follow by placing them within a comparative discussion of recent and contemporary thinking about scale in cartography and geographic information science, physical, and human geography. Whereas we stress here the differences that have recently emerged in conceptualizations of scale within geography, the concluding chapter of the book will seek common ground. As our concluding chapter suggests, these differences are not as stark as they at first seem, but rather are testimony to the richness of a discipline that embraces the human and natural sciences.

Cartographic Scale

Initial thinking on scale paralleled developments in mapmaking. A deterministic method of calculating scale was derived from the idea of the map as a general measurement/storage device (rather than seeing a map as a mechanism to depict some specific distribution – such as the thematic map). As the "science" of cartography emerged in eighteenth-century France with the development of modern geodesy and the creation of the first state-sponsored national map – the Carte de Cassini – the problem of measurement consumed the cartographic community. Large-scale topographic mapping required that the precise relationship between the map and earth be known. It was during this time that formalized state-sponsored cartographic scales were sanctioned, first by the French and then in other European countries. The Carte de Cassini, finished in 1789 on 180 sheets, was published at a representative fraction (RF) of 1:86,400 (and the RF value has now become the standard method for representing scale on maps). An RF value of 1:24,000 indicates that one distance unit on the map represents 24,000 units on the surface of the earth. It is a simple and very functional way to represent scale: the relationship is "unitless," in that any distance measure may be inserted (feet, meters, miles), and also intuitive. However, a serious problem is that any enlargement or reduction of the original map makes the RF value wrong – since it alters the map units without adjusting the earth units. The cartographic graphical scale (the scale bar seen on many maps) is a more reliable representation of scale because it is reduced or enlarged along with the accompanying map. Cartographers have identified three major methods for representing scale on maps: the representative fraction, the graphical scale, and the verbal statement (e.g., one inch equals one mile). All these concepts of scale in cartography, and geographic information science (GISc), are mathematical, with the "representative fraction" being the standard measure.

An interesting development in the cartographic representation of scale is the idea that, within a virtual environment, *there is no scale.* The argument is made in the geographic information science community that many computer databases are "scaleless" in that the traditional concept of scale is not meaningful for electronic data (Goodchild, this volume). Of course, one cannot ignore the fact that most of these databases were acquired from paper maps with an established scale and thus a certain intrinsic "fitness for use." An excellent example of this may be found with the United States Bureau of the Census TIGER (Topologically Integrated Geographically Referenced) files. Much of the geographical detail was geocoded from existing 1:100,000 US Geological Survey maps and thus a "fitness for use" is in "in the vicinity" of 1:100,000. A user would be unwise to map these data at either 1:50,000 or 1:500,000, because the level of geographical detail that was captured in the geocoding process is appropriate at this 1:100,000 scale. Yet the digital data themselves – the strings of *x-y* coordinate pairs stored as binary digits – have no real scale.

While topographical mapping normally occurs at a large cartographic scale (e.g., 1:24,000), much of the thematic, or special purpose, mapping starting in mid- to late-nineteenth-century Europe has been at intermediate or small scales (e.g., 1:500,000). This terminology of large and small is a major source of confusion in the understanding of cartographic scale. Mathematically, a fraction of 1:24,000 is larger than that of 1:500,000, meaning that a detailed map (say, of a village) has a smaller cartographic scale than a map of the world.

Geographers think of scale in the opposite way, however. Human geographers, for example, think about small-scale neighborhood problems and large-scale national problems, meaning that large scale refers to a large area; and small scale to a small area. The scales used by human geographers range from the human body to the globe:

- human body;
- household;
- neighborhood;
- city;
- metropolitan area;
- province/state;
- nation-state;
- continent;
- globe (all adjectives, or all verbs).

These scales have generally been thought of as nested, although the true relationships among scales are often more complicated than this. For example, the Twin Cities metropolitan area is not nested within the state of Minnesota, but stretches into Wisconsin.

A related range of spatial scales, designed for environmental health policy and research, was proposed by Sexton et al. (2002). This includes:

- personal exposure;
- city block / factory;
- city;
- state;
- country / continent;
- earth.

Operational scale

Operational scale refers to the logical scale at which a geographical process takes place (Lam, this volume). For example, the spatial mismatch theory in human geography – i.e., that American inner city residents have lost access to employment as firms moved to the suburbs – operates by definition at the metropolitan scale. In the US, the city scale is too small for examining this theory (because suburbs typically are separate municipalities), whereas the nation-state scale is too large. In the biophysical realm, stream turbulence is studied at the "reach" scale, not for the entire stream or drainage basin. Most biophysical processes operate at particular spatial and temporal scales (see Phillips, this volume), and plenty of examples can also be found in human geography. Gentrification is typically localized to small areas of the inner city, and foreign direct investment occurs at the international scale. Yet the social construction of scale means that there are many other examples where there is no characteristic operational scale (see Smith, this volume).

Spatial resolution

The terms "geographical scale" and "spatial resolution" are often conflated with one another in geography. Whereas the common definition of spatial scale deals with the geographical "extent" of a study area, spatial resolution details the granularity of the data. In nearly all geographic inquiry, it is necessary not only to select the geographical area – or scale – but also the resolution of the data to be analyzed. This is most easily explained with remote sensing data. The study of land-use/land-cover at a particular spatial scale, perhaps the Twin Cities metropolitan region, can involve different spatial resolutions of data. Possibilities include Landsat multispectral scanning imagery (79 m resolution), Thematic Mapper Imagery (30 m resolution), or SPOT imagery (10 m resolution). Each different resolution, or "grain," will likely result in a different empirical result. Similarly, census data can be analyzed at a variety of resolutions (McMaster, Leitner, and Sheppard, 1997, provide an example of how much difference this makes in analyses of environmental equity). These include county, Minor Civil Division (MCD), tract, block group, and block resolutions, and increasingly even a parcel or address level – raising many concerns about privacy. In nearly all of geographic inquiry, one must not only select the geographical area – the scale – but also the resolution of the data to be analyzed. The granularity relates to what is known as the Modifiable Areal Unit Problem (MAUP), discussed below.

Goodchild and Quattrochi (1997) discuss the relationship between scale and resolution. "Geographic scale," they assert:

> is important because it defines the limits to our observations of the earth. All earth observations must have a small linear dimension, defined as the limiting spatial resolution, the size of the smallest observable object, the pixel size, the grain of the photographic emulsion, or some similarly defined parameter. Observation must also have a large linear dimension, defining the geographic extent of the study, project, or data collection effort. There are many ways of defining both parameters, and this is one of the factors contributing to the richness of the scale issue. (Goodchild and Quattrochi, 1997: 2)

This statement relates to our own argument, where nearly all studies require a grain or resolution – the small linear dimension – and a geographic extent – or large linear dimension.

Modifiable Areal Unit Problem (MAUP)

Not surprisingly, geographers (and others) have discovered that the resolution, or grain, of the analysis can affect geographic inquiry. An analysis of poverty at the census block level, of course, will likely yield different results than at the block-group, tract, or MCD (Minor Civil Division) levels. Likewise a land-use/land-cover analysis using 10 m resolution data will likely lead to a different classification than that using remote sensing imagery at a 30 m resolution. Furthermore, as discussed later, even for a particular resolution different spatial units (e.g., different ways of grouping blocks into block groups) can result in remarkably different empirical findings (Openshaw and Taylor, 1979). Although this "discovery" seems rather intuitive, it implies that the geographic analyst faces great problems in identifying which resolution is best, or even optimal. Walsh et al. (this volume) address this very issue, and there is a growing literature on MAUP, and its effect on various types of geographical analyses.

Cartographic generalization

Cartographers have worked for centuries on determining the appropriate amount of information to include on maps of different scales. The amount of information possible at a scale of, for instance, 1:24,000, is different than the information possible at 1:500,000. The filtering of information at one scale to create a map at a smaller scale is known as cartographic generalization. For example, Hudson (1992) identifies the effect of scale on what can be depicted in a map of 5 by 7 inches:

- a house at a scale of 1:100;
- a city block at a scale of 1:1,000;
- an urban neighborhood at a scale of 1:10,000;
- a small city at a scale of 1:100,000;
- a large metropolitan area at a scale of 1:1,000,000;
- several states at a scale of 1:10,000,000;
- most of a hemisphere at a scale of 1:100,000,000;
- the entire world with plenty of room to spare at a scale of 1:1,000,000,000.

Hudson (1992: 282) explains, "These examples, ranging from largest ($1{:}10^2$) to smallest ($1{:}10^9$) scale, span eight orders of magnitude and, as a practical matter, cover the spectrum of scales at which geographers are likely to use maps."

These differences pose the problem known as cartographic generalization: How should information on a map be simplified, or filtered, when it is redrawn at a smaller cartographic scale? The European cartographic community became aware of the generalization problem in the early part of the twentieth century. In a 1908 *Bulletin* of the American Geographical Society, the German cartographer, Max Eckert, wrote:

> In generalizing lies the difficulty of scientific map-making, for it no longer allows the cartographer to rely merely on objective facts, but requires him *(sic)* to interpret them subjectively. To be sure the selection of the subject matter is controlled by considerations regarding its suitability and value, but the

> manner in which this material is to be rendered graphically depends on personal and subjective feeling. But the latter must not predominate: the dictates of science will prevent any erratic flight of imagination and impart to the map a fundamentally objective character in spite of all subjective impulses. It is in this respect that maps are distinguished from fine products of art. Generalized maps and, in fact, all abstract maps should, therefore, be products of art clarified by science. (Eckert, 1908: 347)

By the 1970s, cartographers were hard at work attempting to "discover" a theory of map generalization, and had identified a set of fundamental elements, including selection, simplification, classification, and symbolization (Robinson, Sale, and Morrison, 1978). Selection, often considered a prior step to generalization, involves identifying which classes of features to retain. For instance, does one retain or eliminate water bodies, transportation networks, and/or vegetation on the generalized map? Simplification involves determining the important characteristics of the data, the retention and possible exaggeration of these characteristics, and the elimination of unwanted detail. Classification is identified as the ordering or scaling and grouping of data, while symbolization defines the process of graphically-encoding these scaled/grouped characteristics (McMaster and Shea, 1992).

Unfortunately, a process (generalization) that was reasonably well-understood for paper maps, and even codified in certain instances by agencies such as the United States Geological Survey, has become a classical "ill-defined" problem in the digital domain. The problem is how to identify the appropriate "techniques," or computer algorithms, to accomplish what had been for many centuries a manual process (for further detail, see McMaster and Shea, 1992; Buttenfield and McMaster, 1991). Figure Intro.1 depicts original and generalized versions of the census tracts for Hennepin County, Minnesota. The generalized version represents a simplification, where coordinate pairs that were deemed "redundant" (superfluous) for representing the shape of the line have been eliminated.

Scale in Biophysical Geography

Biophysical geographers, like earth scientists more generally, seek to account for the spatial dynamics of complex ecological, meteorological, climatic, and geomorphic systems. As the essays in this book illustrate, biophysical geography differs from the earth sciences in giving more attention to human–environment relations. Yet they share a concern for constructing general explanations, rooted in positivist or realist philosophies, and for integrating temporal and geographic scale (Rhoads and Thorn, 1996; Richards, Brooks,

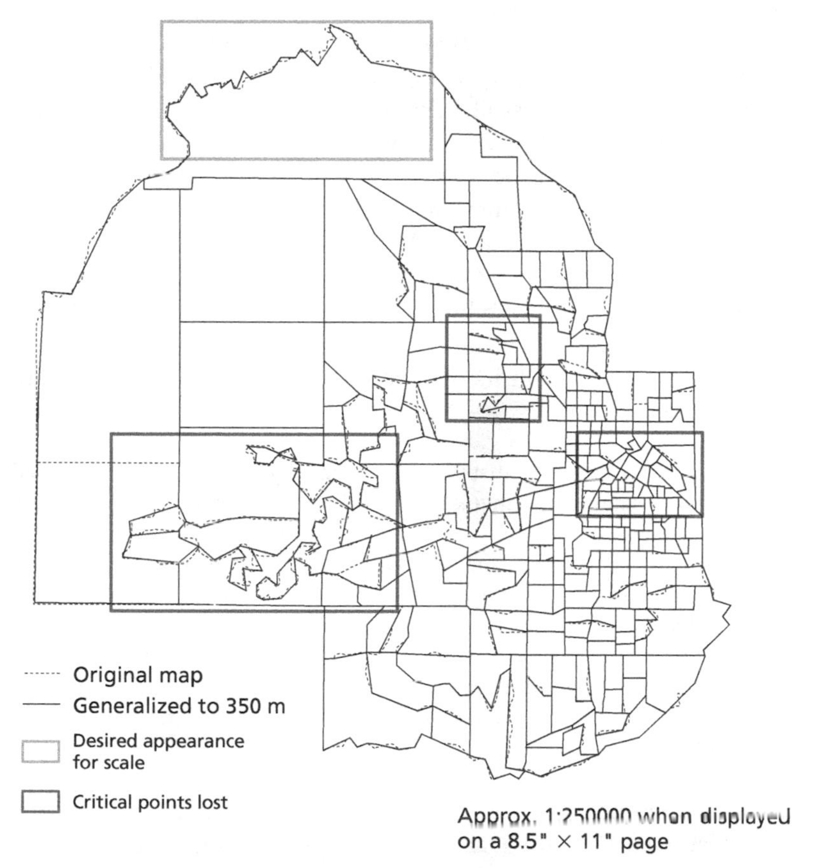

Figure Intro. 1 *Effects of uniform generalization: Douglas–Peucker Algorithm.*

Clifford, Harris, and Lane, 1997; Phillips, 1999). Jonathan Phillips (1997: 99) argues that research on scale in the earth sciences addresses four kinds of issues:

- *identifying and measuring* the range of spatial and temporal scales, and the *characteristic* (operational) *scale* of particular processes;
- *reconciling* the scales of processes with those of observation and measurement;
- issues of *dimensionality and similarity* – addressing ranges of scales across which relationships are constant, or where straightforward rules for down- or up-scaling can be derived;
- operational problems of *scale linkage* – carrying out cross-scale analysis, in situations where relationships vary across scales (*multiscale analysis*, Wilbanks, 2001).

This list indicates the importance of keeping in mind a multiplicity of temporal and spatial scales. Biophysical phenomena vary from highly localized and very fast processes, such as stream eddies of air turbulence, to very large-scale long-term process like climate change. This can pose significant challenges for mapping (Ziegler et al., this volume), and analyzing (Phillips, this volume) biophysical processes. As Bauer, Veblen, and Winkler (1999: 681) note, developing explanations at these very different temporal and spatial scales poses distinct methodological and philosophical problems: "A fundamental question to be addressed in this regard is whether there is an irreducible incommensurability to nature when viewed and described at different scale levels. . . . Does this imply that there may exist different levels of understanding, each with its own complexities and fundamental laws?" They continue:

> At the smallest scales, scientists have tended to favor concisely expressed, deterministic relations that invoke force balances or conservation principles (of mass, momentum, vorticity, entropy, or energy). At intermediate scales, much of the mathematical formality is retained, but some of the deterministic physics or chemistry are replaced by parameterizations that invoke constitutive coefficients or various other coefficients of bulk behavior... At the largest scales, descriptions of system behavior often assume probabilistic properties or an idiographic and historical character, although exceptions exist (e.g., General Circulation Models). (Bauer, Veblen and Winkler, 1999: 682)

In trying to simplify cross-scale analysis, physical geographers turn to hierarchy theory (see Easterling and Polsky, Phillips, and Walsh et al., this volume).

Hierarchy theory

In trying to make sense of this complexity, biophysical geographers often turn to hierarchy theory – an idea pioneered in ecology (Levin, 1992). According to hierarchy theory, nature subdivides itself into a hierarchical system with both a vertical structure of levels, and a horizontal structure of "holons" (Figure Into.2a). Holon derives from the Greek for whole (*holos*) and part (*on*), conveying the idea that subsystems at any level act as wholes with respect to lower levels of the hierarchy, but are parts of units at higher levels. By definition, interactions are significantly stronger both within holons than between holons at a particular level, and within a level rather that across the "surfaces" separating levels. Furthermore, each level can be distinguished from others by its time and space scale: Processes at lower

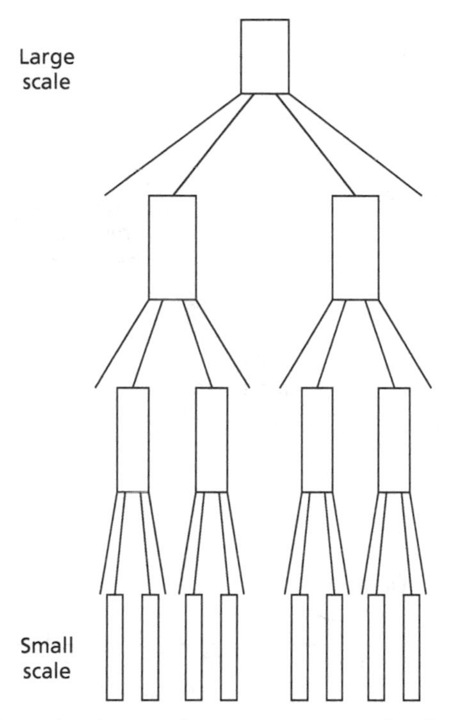

Figure Intro. 2a *Hierarchy theory: the arrangements of scales and spatial units.*

levels occur both more rapidly and across smaller spatial scales than those at higher levels.

Hierarchy theory has several implications for scale and geographic inquiry in biophysical geography. First, natural phenomena can be separated according to distinctive time and space scales (Figure Intro.2b). Second, it follows from this that different processes can indeed be expected to have characteristic spatiotemporal scales at which they operate – the previously mentioned "operational scale." Third, this implies that multiscalar analysis can be dramatically simplified (indeed Wu, 1999, argues that without hierarchy theory little simplification is possible). In particular, when analyzing any particular level the processes operating at the next higher scale can be regarded as constraints. They are so much slower, and show so little spatial variation at the scale of analysis, that they can be treated as constants. Processes operating at the next lower scale are conceptualized as driving change at the scale of interest, but run so much more quickly that they can be regarded as having reached an equilibrium state. This means that they can be approximated as fixed initial conditions for the purposes of modeling change at the scale being studied. Wu concludes that hierarchy theory

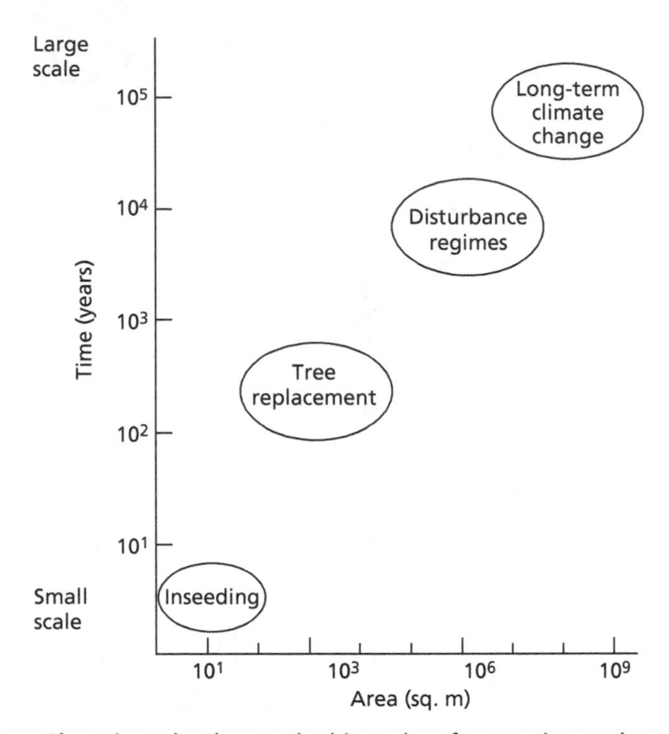

Figure Intro. 2b *Hierarchy theory: the hierarchy of space–time scales.*
Source: Drawn by authors, based on Haila and Levins (1992), Goodchild and Quattrochi (1997), Wilbanks (2001), and Wu (1999).

(i) provides a common approach to scientific analysis applicable at all scales of geographic interest, (ii) makes it possible to restrict analysis to just three scales at once, and (iii) makes it possible to either simplify middle-number systems of organized complexity to small-number systems of organized simplicity, or at least reduce the complexity of a system to "a more manageable level" (1999: 6). This last issue is addressed by Easterling and Polsky (this volume).

Hierarchy theory simplifies analysis, because it assumes that spatial and temporal scale covary. Thus the four issues listed by Phillips (1997) can treat spatial and temporal scale simultaneously. Not all physical geographers accept hierarchy theory as a complete description of nature, however. Bauer and his colleagues provide a summary diagram depicting three strategies for studying spatial and temporal scale in biophysical geography that takes into account a more complex relationship between temporal and spatial scale (Figure Intro.3). The vertical line represents studies of spatial pattern across a range of spatial scales for a particular temporal scale, perhaps a multiscale drainage basin analysis (in their view, traditionally

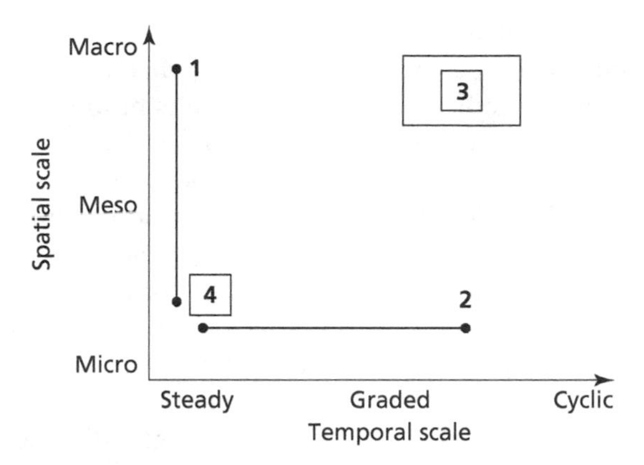

Figure Intro. 3 *A space–time diagram*
Source: Redrawn from Bauer et al. (1999: 684)

involving map analysis). The horizontal line represents studies of a range of temporal scales for a particular scaled study area, such as the analysis of weekly, monthly, and yearly meteorological data at the Twin Cities International Airport. They also note that the scaling-up indicated by these two lines, often assumed to be seamless, is highly problematic. The squares and rectangles on the diagram represent studies examining a restricted range of spatial and temporal scales. Hierarchy theory provides a justification for such restrictions, as does the idea that particular biophysical processes have a characteristic operational space–time scale.

Scale in Human Geography

In human geography, thinking on the role and nature of geographical scale has undergone a particularly fundamental shift during the last fifteen years, raising some questions about scale that have received little attention elsewhere in geography.

Ecological inference, scale-variance analysis and the Modifiable Areal Unit Problem

Human geographers have long been aware that choice of scale and resolution affects empirical analysis. Statisticians and sociologists already identified confounding effects of scale on spatial analysis prior to the emergence of the spatial science tradition in human geography in the 1960s. They observed a tendency for the correlations between two variables to increase for spatial units of larger resolution, and on that basis identified the problem of the ecological fallacy (Gehlke and Biehl, 1934; Robinson, 1950). This resulted in identification of the ecological fallacy; i.e., that is erroneous to infer relationships between individuals from the results of more aggregate-scale studies. During the 1960s and 1970s, human and physical geographers attempted to develop and use statistical methods that would take into account the impact of scale on the relations between variables. Two broad approaches were developed: Spectral and Fourier analysis, and scale-variance analysis. The former approach, adopted from time series analysis, sought to determine the operational spatial and/or temporal scale of a geographic process. Taking advantage of the mathematical properties of Fourier transforms (their ability to partition any spatial or temporal distribution into a series of independent waves of differing wavelength, direction, and amplitude), it became possible to determine which wavelengths were particularly important to describing the spatial distribution of a phenomenon. These were then inferred to be the characteristic or operational scale for that phenomenon (Rayner, 1971; Bennett, 1979). The absence of large continuously distributed spatial data sets precluded extensive application of this approach by human geographers, but they did take an interest in the spatial variability of cyclical time series, particularly unemployment. An extensive body of research studied the characteristic temporal scales of unemployment, and their spatial variation, seeking to measure the transmission of unemployment cycles across space, and whether certain locations characteristically play a leading or lagging role in this diffusion process (for a review see Martin and Spence, 1981).

Scale-variance analysis sought to approximate spectral analysis for smaller spatial data sets collected in discrete space, developing methods for measuring the variance of a single variable at different spatial scales, and for calculating correlations between variables at these different scales (Tobler, 1969; Curry, 1972; Moellering and Tobler, 1972). Here, spatial scales are defined as nested sets of spatial units of different spatial resolution (e.g., neighborhoods nested within cities, nested in turn within metropolitan areas). Examination by Stan Openshaw and Peter Taylor (1979) of the effects of scale and resolution on statistical analysis in human geography led to identification of the modifiable area unit problem, as discussed previously. They showed that correlations computed between pairs of variables varied dramatically in size and in direction (positive or negative) for the same data set, not only when analyzed at different scales but also when analyzed at a constant scale but for different aggregations of more finely grained spatial units. This has become a widely discussed puzzle in GIS and quantitative human geography, and also more generally in the social sciences where its relationship to the ecological fallacy has recently received attention (King, 1997; Johnston and Pattie, 2001).

The construction of space and scale

Problems of resolution, scale, and the stability and predictability of quantitative relationships remain an ongoing theme across quantitative geography, as a number of the chapters in this book indicate. Yet, as the balance of contributions in this book also indicates, human geographic research on scale has taken a very different turn in the last decade. This is a consequence of theoretical and methodological shifts in human geography. Theoretically, more and more human geographers have come to question the adequacy of Euclidean coordinate systems as a way of representing space and time. They are not concerned with the fact that the earth's surface is more accurately represented by spherical than Cartesian coordinates, but with the question of what distance means in the social realm. Human geographers agree that the actual distance between two places may have little to do with the miles separating them. Black and white neighborhoods may only be across the street from one another. Yet the effective social distance separating them, as indicated by minimal social interaction between them, can be enormous. Similarly, although New York and London are far apart in Euclidean space, in other senses, as measured by flows of money, people, and information, they are much closer to one another than either is to other cities a few miles away. Beginning in the 1970s, political economic geographers began to go beyond this, to argue that space, and, by implication, scale, is produced through the characteristic political-economic processes of a certain societal system, such as capitalism (Lefebvre, 1974 {1991}; Smith, 1984; Harvey, 1996). Although the difference between this position and the alleged "spatial fetishism" of quantitative human geography was exaggerated (Sheppard, 1995), a vital question had been raised. If space is not an exogenous and fixed dimension, but is shaped by societal processes, how can we account for the construction of space, and scale, in our explanations of society?

This has become the central concern of scale theorists in contemporary human geography. Their analysis begins with two claims: First, if space is socially constructed, the same must also be true for scale, so we need to think about how scales come into existence. Second, if scale is socially constructed, then we cannot simply take for granted the existence and importance of the geographic scales usually invoked in human geographic writing: Neighborhood, city, regional, national, and global. Rather, we need to understand not only why their relative importance may vary over space and time, but also whether these are even the right scales to be thinking about.

Economic, political, technological, and discursive constructions of scale

A lively debate exists about how and why scales are created, and become more, or less, important, discussing the relative importance of economic, political, social, and gendered processes and representations. The strongest tradition, emerging out of Marxist-inspired political economic theories of the space economy, stresses the *production* of scale (see Smith, Swyngedouw, this volume); i.e., how economic and political processes shape the emergence of scale. Processes include the geographic strategies of capitalist firms, of political institutions such as the nation-state, and of labor organizing to improve livelihood conditions in the face of challenges posed by capital mobility and/or state strategies. Neil Smith and Ward Dennis (Smith and Dennis, 1987) argued that the greatest variability in uneven development within the United States shifted during the 1970s from a regional (sunbelt-snowbelt) scale to more localized scales, as a result of an increased emphasis on interurban and interstate competition for investment, and a devolution of the responsibility for achieving and maintaining economic prosperity from the Federal to the US state- and local scales.

Erik Swyngedouw identified what he dubbed "glocalization:" political economic forces driving globalization that are simultaneously making both the global scale and also subnational metropolitan regions more important scales in the geography of economic change, whereas the national scale is becoming less important (Swyngedouw, 1997a; 1997b). Neil Brenner has gone on to argue that these scalar shifts are in fact one of the most important distinguishing features of contemporary globalization (Brenner, 1999). Between World War II and the early 1970s, economic prosperity in the first world was a consequence of strong and effective Fordist national policies coordinating domestic demand with supply, ensuring good wages for organized labor, and controlling the movement of capital, commodities, and people across national boundaries. Since then, nation-states have lost or abrogated many of their powers to control economic development at the national scale, thereby underwriting glocalization. Andrew Herod, among others, has reported on a variety of case studies demonstrating that organized labor has also played an important role in shaping the scales at which uneven development occurs, ranging from support for global geopolitical strategies by the US government, to fighting corporate strategies to move jobs abroad, to employing scalar strategies themselves in labor organizing on the East Coast of the United States (Herod, 1998; 2001). More abstractly, David Harvey recently has begun to frame the whole theory of uneven development as a consequence of the production of scale, at scales ranging from the body to the globe (Harvey, 2000).

Others have placed much more influence on the *political construction* of scale (Delaney and Leitner, 1996). In this view, the politics of scale is not just driven by economic dynamics but is politically constructed, through political movements, institutions, and networks whose activities are not reducible to political economic processes (Leitner, this volume). This approach has been used to examine, for example, the emergence of a supranational regime for regulating immigration to the European Union, the strategies of US peace movements in the 1980s, the geography of Italian political change, and the emergence of political networks transcending the hierarchical territorial spaces controlled by supranational, national, and subnational states (Agnew, 1997; Leitner, 1997; Miller, 1997; Leitner, Pavlik, and Sheppard, 2002). The politics of scale has also been an important theme in the work of Neil Smith (1992; 1996) and Swyngedouw. Smith (this volume) developed the concept of scale jumping to describe the strategies that effective social movements must develop in order to take their concerns beyond the local level. Harvey (2000) elaborates at length on this theme, arguing that political movements must be willing to operate simultaneously, and even in mutually contradictory ways, at different scales if they are to effect social change. Swyngedouw (this volume) argues that the scales created in struggles over access to and management of water in Ecuador and Spain create spaces where a progressive politics of scale, in opposition to the authoritarian tendencies of those dominating larger scales, can be articulated. His research applies a constructionist approach to scale to focus on human environment relations. Many human geographers argue that political economic processes not only produce space, but also produce nature (Smith, 1984; Castree, 1995). Swyngedouw's application of a constructionist scale theory to the natural environment draws on this approach, creating an interesting contrast to other analyses of scale and human environment relations (see Easterling and Polsky, Walsh et al., Lam this volume).

The impact of changing communications and information technologies on the spatial organization of society is now an important research area in geography, and can the extended to examine how technological change, itself rooted in a variety of societal processes, is shaping scale. Goodchild (this volume) suggests, for example, that geographically mobile information technologies are making scale, particularly local scales, a more important attribute of geographic information.

In order to open up constructionist approaches to scale to a gendered analysis, it is necessary to extend the range of scales downward to embrace the home and the body – scales that are central to a feminist approach – and to open up the question of the agency of women in the construction of scale, much as Leitner and Herod argue, respectively, for including the agency of social movements and organized labor in accounts of the construction of scale (Leitner, 1997; Herod, 2001). Sallie Marston (2000), argues that the approaches described above systematically neglect the role of social reproduction in constructing scale. She appropriately highlights this as another case where unwaged household activities are left out of political and economic analyses, and argues that American feminist movements were influential at the turn of the nineteenth century in shifting responsibility for social welfare and collective consumption upwards from the scale of the household to those of the Federal and local states (Marston, this volume). She also suggests that this success may itself lie behind the subsequent crisis in collective consumption that helped bring about an end to Fordism in the 1970s, and a return to emphasizing economic development, rather than welfare and redistribution, at the Federal and local state levels. Similarly, Linda McDowell (2001) is beginning to frame the complex connections between bodies, organizations, and the economy in terms of geographic scale.

Discursive representations of scale are just beginning to emerge as a distinct theme, drawing on the more general concern, during the recent cultural turn in human geography, with discourse theory. In discourse theory, it is argued that rather than language reflecting the world, the world comes to reflect language. We do not simply observe the world and note what is objectively important and accurate. Rather, the very observations we make, and the practices we pursue, depend on how we think about the world. In turn, our thoughts and beliefs are shaped by the discourses we are exposed to. In this view, the importance of certain ideas and phenomena, such as certain scales, and the ways in which they are routinely perceived and thought about, are shaped by societal discourses. Kelly (1999) stresses that an important element of the politics of scale is discursive; certain scales become important as a result of discourses highlighting them. Because human practices reflect discourse this in turn also enhances their material importance. For example, globalization has emerged as a dominant term in social discourse, and influential discourses about globalization represent it in a certain way – as an inevitable force that, in General Colin Powell's words, benefits everyone. When such tropes dominate social discourse, they shape beliefs about both the pervasiveness and nature of globalization. If we buy into such discourses about the importance of globalization, then we are more likely to interpret our observations about the world in a way that prioritizes globalization, thereby pursuing practices that reinforce its pervasiveness and inevitability.

Although the relative merits of, and relations among, these different perspectives on the construction of scale are still the subject of lively debate (see Leitner, Marston, this volume), there is consensus on the need to move away from thinking about geographic scales as pregiven dimensions of society, to thinking about their social construction. This consensus challenges at least four aspects of more conventional thinking about scale. First, it is argued that the significance of the different scales usually discussed, from the body to the globe, is not fixed or pre-determined. Their relative importance, and even their existence, depends on societal processes. Second, this consensus challenges the notion that scales are hierarchical – that larger scales shape what happens at lower scales (Marston, this volume). For example, initially highly localized social movements, such as the Zapatistas in Chiapas province, Mexico, have been able to jump scale and influence global processes. Third, this consensus challenges the idea that scales are nested – that the spatial units constituting a lower geographic scale are fully contained within the spatial units constituting higher geographic scales. It is noted, for example, that cities can develop their own "foreign policies" and seek help from outside the region or nation where they are located, without seeking permission from institutions representing the higher scales within which they are found (Smith, this volume). Fourth, this consensus challenges the idea that geographic scales are subdivided into spatially contiguous units of a certain size (or resolution).

Networks and scale

As Peter Taylor (this volume) and Helga Leitner (this volume) both stress, consideration of networks, linking places together across space, raise further, and these questions about geographic scale. Interurban networks are scaled geographic entities that constitute a new scale, and these cannot be reduced to such other conventional scales as nations or regions. They are not hierarchical, nested, or spatially contiguous units, but stretch across the hierarchical, nested spaces of political geography – spanning space instead of covering it (Leitner, Pavlik, and Sheppard, 2002). Indeed, consideration of networks shows clearly the importance of continually critically assessing and revising our conceptions of scale in human geography, instead of trying to capture scale through some fixed definitions. This is always hard, as rethinking always begins by drawing on familiar terms that themselves make it more difficult to imagine alternatives. For example, those theorizing the construction of scale often draw on, and use as examples in their own work, conventional scalar units of political geography: Neighborhood, city, region, nation, and supranational blocs, and the globe. These are hierarchically nested territories with well-defined boundaries, thus typically exhibiting all the characteristics of scale that these theorists seek to call into question.

The challenge, here as more generally in geography, is to avoid the trap of preexisting concepts, retaining the kind of openness and critical awareness that takes our thinking forward in a rigorous but reflective manner.

References

Agnew, J. A. 1997: The dramaturgy of horizons: Geographical scale in the "Reconstruction of Italy" by the new Italian political parties, 1992–95. *Political Geography*, 16(2): 99–121.

Bauer, B., T. Veblen, and J. Winkler 1999: Old methodological sneakers: Fashion and function in a cross-training era. *Annals of the Association of American Geographers*, 89(4): 679–86.

Bennett, R. J. 1979: *Spatial Time Series*. London: Pion.

Brenner, N. 1999: Beyond state-centrism? Space, territoriality, and geographical scale in globalization studies. *Theory and Society*, 28(1): 39–78.

Buttenfield, B. P. and R. B. McMaster (eds) 1991: *Map Generalization: Making Rules for Knowledge Representation*. London: Longman.

Castree, N. 1995: The nature of produced nature. *Antipode*, 27(1): 12–48.

Carroll, L. 1893: *Sylvie and Bruno concluded*. [New York: Dover Publications, 1988.]

Curry, L. 1972: A bivariate spatial regression operator. *The Canadian Geographer*, 16(1): 1–14.

Delaney, D. and H. Leitner 1996: The political construction of scale. *Political Geography*, 16(2): 93–7.

Eckert, M. 1908: On the Nature of Maps, and Map Logic. Translated by W. Joerg. *Bulletin of the American Geographical Society*, 40(6): 344–51

Gehlke, C. E. and K. Biehl 1934: Certain effects of grouping upon the size of the correlation coefficient in census tract material. *Journal of the American Statistical Association*, 29: 169–70.

Goodchild, M. and D. Quattrochi 1997: Introduction: Scale, multiscaling, remote sensing, and GIS. In D. Quattrochi and M. Goodchild (eds), *Scale in Remote Sensing and GIS*, New York: Lewis Publishers.

Haila, Y. and R. Levins 1992: *Humanity and Nature: Ecology, Science, and Society*. London: Pluto Press.

Harvey, D. 1996: *Justice, Nature and the Geography of Difference*. Oxford: Blackwell Publishers.

Harvey, D. 2000: *Spaces of Hope*. Berkeley: Berkeley, CA: University of California Press.

Herod, A., (ed.) 1998: *Organizing the Landscape*. Minneapolis, MN: University of Minnesota Press.

Herod, A. 2001: *Labor Geographies: Workers and the Landscapes of Capitalism*. New York: Guilford.

Hudson, J. C. 1992. Scale in space and time. In Ronald Abler, Melvin Marcus, and Judy Olson (eds) *Geography's Inner Worlds*. New Brunswick, NJ: Rutgers University Press, 280–300.

Johnston, R. J. and C. Pattie 2001: On geographers and ecological inference. *Annals of the Association of American Geographers*, 91(2): 281–2.

Kelly, P. F. 1999: The geographies and politics of globalization. *Progress in Human Geography*, 23(3): 379–400.

King, G. 1997: *A Solution to the Ecologicial Inference Problem: Reconstructing Individual Behavior from Aggregate Data*. Princeton, NJ: Princeton University Press.

Lefebvre, H. 1974 {1991}: *The Production of Space*. Oxford: Blackwell Publishers.

Leitner, H. 1997: Reconfiguring the spatiality of power: the construction of a supranational migration framework for the European Union. *Political Geography*, 15(2): 123–43.

Leitner, H., C. Pavlik, and E. Sheppard 2002: Networks, governance and the politics of scale: Inter-urban networks and the European Union. In A. Herod and M. Wright (eds), *Geographies of Power: Placing Scale*, Oxford: Blackwell, 274–303.

Levin, S. A. 1992: The problem of pattern and scale in ecology. *Ecology*, 73(6): 1943–67.

Marston, S. 2000: The social construction of scale. *Progress in Human Geography*, 24(2): 219–42.

Martin, R. L. and N. Spence 1981: Economic geography. In N. Wrigley and R. J. Bennett (eds), *Quantitative Geography: A British View*. London: Routledge and Kegan Paul, 322–34.

McDowell, L. 2001: Linking scales: or how research about gender and organizations raises new issues for economic geography. *Journal of Economic Geography*, 1(2): 227–50.

McMaster, R.B. and K.S. Shea. 1992: *Generalization in Digital Cartography*. Resource Publications in Geography. Washington, DC: Association of America Geographers.

McMaster, R. J., H. Leitner, and E. Sheppard 1997: GIS-based environmental equity and risk assessment: Methodological problems and prospects. *Cartography and Geographic Information Systems*, 24(3): 172–89.

Meyer, William B., Derek Gregory, B.L. Turner, and Patricia McDowell. 1992: The Local-Global Continuum. In Ronald Abler, Melvin Marcus, and Judy Olson (eds) *Geography's Inner Worlds*. New Brunswick, NJ: Rutgers University Press, 255–79.

Miller, B. 1997: Political action and the geography of defense investment: geographical scale and the representation of the Massachussetts miracle. *Political Geography*, 16(2): 171–5.

Moellering, H. and W. Tobler 1972: Geographical variances. *Geographical Analysis*, 4(1): 34–50.

Openshaw, S. and P. J. Taylor 1979: A million or so correlation coefficients: Three experiments in the modifiable areal unit problem. In N. Wrigley (ed.), *Statistical Applications in the Spatial Sciences*. London: Pion, 127–44.

Phillips, J. D. 1997: Humans as geological agents and the question of scale. *American Journal of Science*, 297(1): 98–115.
Phillips, J. D. 1999: *Earth surface systems: complexity, order and scale*. Malden, MA: Blackwell Publishers.
Rayner, J. N. 1971: *An Introduction to Spectral Analysis*. London: Pion.
Rhoads, B. L. and C. E. Thorn, (eds) 1996: *The Scientific Nature of Geomorphology*. Chichester, UK: John Wiley.
Richards, K. S., S. M. Brooks, N. J. Clifford, et al. 1997: Theory, measurement and testing in 'real' geomorphology and physical geography. In D. R. Stoddard (ed.), *Process and Form in Geomorphology*. Oxford: Blackwell Publishers, 265–92.
Robinson, W. S. 1950: Ecological correlations and the behavior of individuals. *American Sociological Review*, 15: 351–7.
Robinson, A.H. and Barbara Bartz Petchenik 1976: *The Nature of Maps: Essays Towards Understanding Maps and Mapping*. Chicago, IL: University of Chicago Press.
Sexton, K., L. A. Waller, R .B. McMaster, G. Maldonado, and J. L. Adgate. 2002. The Importance of spatial effects for environmental health policy and research. *Human and Ecological Risk Assessment* , 8(1): 109–25.
Sheppard, E. 1995: Dissenting from spatial analysis. *Urban Geography*, 16: 283–303.
Smith, N. 1984: *Uneven Development: Nature, Capital and the Production of Space*. Oxford: Blackwell Publishers.
Smith, N. 1992: Geography, difference and the politics of scale. In J. Doherty, E. Graham and M. Malek (eds), *Postmodernism and the Social Sciences*. London: Macmillan, 57–79.
Smith, N. 1996: Spaces of vulnerability: The space of flows and the politics of scale. *Critique of Anthropology*, 16(1): 63–77.
Smith, N. and W. Dennis 1987: The restructuring of geographical scale: coalescence and fragmentation of the northern core region. *Economic Geography*, 63(2): 160–82.
Swyngedouw, E. 1997a: Excluding the other: The production of scale and scaled politics. In R. Lee and J. Wills (eds), *Geographies of economies*. London: Arnold, 167–76.
Swyngedouw, E. 1997b: Neither global nor local: "Glocalization" and the politics of scale. In K. Cox (ed.), *Spaces of Globalization: Reasserting the Lower of the Local*, New York: Guilford, 137–66.
Tobler, W. 1969: Geographical filters and their inverses. *Geographical Analysis*, 1: 234–53.
Wilbanks, T. 2001: Geographic scaling issues in integrated assessments of climate change. *Integrated Assessment*, 3(2–3): 100–14.
Wu, J. 1999: Hierarchy and scaling: Extrapolating information along a scaling ladder. *Canadian Journal of Remote Sensing*, 25(4): 367–80.

1 Fractals and Scale in Environmental Assessment and Monitoring

Nina Siu-Ngan Lam

Introduction

Broadly defined, scale has four meanings: the spatial extent of a study, its data or image resolution, the spatial extent of a spatial process, and its representation through a map (Lam and Quattrochi, 1992; Cao and Lam, 1997). While we will present a more precise definition of scale later in this chapter, it suffices to say that scale has been a fundamental problem in geography and in many disciplines that utilize spatial data. In geography, there is a long history of literature that addresses the issues of scale (e.g., Harvey, 1968; Stone, 1972; Watson, 1978). The famous Modifiable Areal Unit Problem (MAUP), which states that results derived from data collected at smaller areal units could be very different from those collected at larger areal units, highlights some of the key issues of scale (Openshaw, 1984). In other disciplines, such as ecology, the fundamental concept of ecological diversity is dependent on the scale at which it was defined and measured. A landscape pattern may look spatially homogeneous at one scale but heterogeneous at another (Turner et al., 1989; O'Neill et al., 1989). It has been argued that scale is as fundamental as time and space and therefore should be considered as a basic dimension of digital geographic data (Quattrochi and Goodchild, 1997).

Scale is important because it contributes to the uncertainties of many scientific studies, and this applies to all four definitions of scale. Scale is known to affect the formulation of a study, its information content, its analysis methods, interpretations of its findings, and henceforth conclusions about its patterns and underlying processes. In remote sensing and GIS, renewed interest in the study of scale is evident from a significant increase in literature on scale in the last two decades (Quattrochi and Goodchild, 1997; Tate and Atkinson, 2001). These scale-related studies generally focus in three areas: the investigation and simulation of the scale effects on the analysis of a phenomenon or landscape indices (Nellis and Briggs, 1989; Hou, 1998); the search for optimal scales (Moellering and Tobler, 1972; Woodcock and Strahler, 1987; Cao and Lam, 1997); and the development of new methods that can accommodate or reflect the nature of scale, such as fractals, variograms, and wavelets (Lam and Quattrochi, 1992; Myers, 1997; Zhu and Yang, 1998; Atkinson and Tate, 1999). While significant advances toward the study of scale in remote sensing and GIS have been made through these various studies, they have not been able to fully measure scale effects and develop strategies to deal with them.

Scale is a difficult problem because its effects are often subtle and its concepts are abstract and confusing. Moreover, the lack of uniformity in terminology, the difference in the nature of the data, and the difference in scope and type of study have made the scale issue seemingly intractable. In this information age, when voluminous data come in all forms and scales and when uncertainties due to scale become a very practical policy issue, a more consorted research effort into scale and scale effects is vital. The University Consortium on Geographic Information Science (UCGIS) has considered scale as a research priority and suggested research themes on scale that are of short-term as well as long-term implications (UCGIS, 1998). Among the various suggestions, the UCGIS calls for the need to develop consistent definitions of scale as well as effective spatial techniques for assessing and characterizing the scale effects.

This chapter focuses on the problem of environmental change detection and monitoring, in which scale plays an essential role. Concomitant with the recommendations made by the UCGIS, our long-term research goal is to develop effective spatial techniques for assessing and characterizing the scale effects in environmental monitoring studies. Recognizing that scale effects exist and can never be eliminated, our research aims at finding a pragmatic solution for mitigating the scale effects by both characterizing and utilizing the scaling properties of the phenomenon in our analysis. Specifically, this chapter introduces the fractal method as a scale-based method for characterizing the type and complexity of landscape patterns, especially those manifested in remote sensing imagery.

The chapter is organized as follows. A definition of scale is first provided to avoid confusion with terminology use in this chapter and to advocate the practice of a standardized use of terminology. This is followed by a discussion of the scale-dependent nature of environmental monitoring and change detection studies. The fractal method is then introduced, and its uses in exploring the scale issues in environmental applications are demonstrated by three examples. First, fractals can be used to characterize spatial patterns at different scale ranges. Second, fractals can be used to demonstrate the complex relationships between landscape, landscape complexity, and scale (pixel resolution and spatial extent). Finally, by changing the spatial extent (observational scale) that simulates a hierarchical or multiscale approach, fractals can be applied both globally (large extent) and locally (small extent) for more accurate change detection and monitoring.

Definitions of Scale

The term scale has a variety of meanings. This chapter adopts the definitions outlined in Lam and Quattrochi (1992) and Cao and Lam (1997), with a focus on examples in the spatial domain. The same meanings of scale can be easily extended to the temporal domain (Figure 1.1).

Within the spatial domain, four common uses of the term "scale" can be identified.

1 The *cartographic* or map scale refers to the ratio between the measurements on a map and the actual measurements on the ground. A large-scale map covers a smaller area and the map generally has more detailed information. However, a small-scale map covers a larger area and the map often has less detailed information about the area.
2 The *observational* or geographic scale refers to the spatial extent of the study or the area of coverage. Under this usage, a large-scale study covers a larger study area, as opposed to a small-scale study that encompasses a smaller study area. For example, a study of global temperature change is a large-scale study that will include most of the earth, compared with a study of urban heat-island effects in a city that focuses only on a specific urban area.
3 The *measurement* scale, or commonly called *resolution*, refers to the smallest distinguishable parts of an object (Tobler, 1988), such as pixels

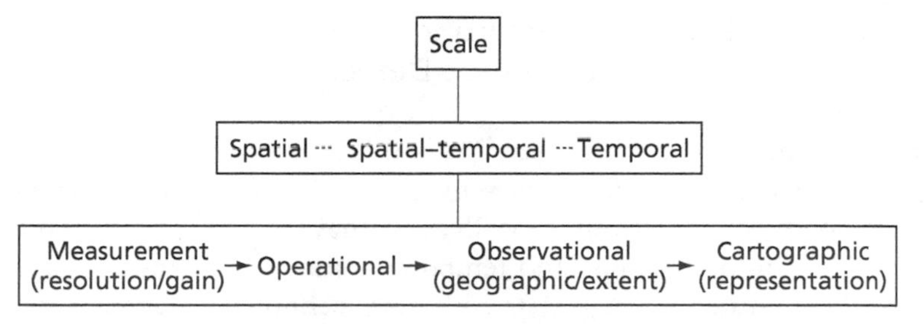

Figure 1.1 *The four meanings of scale.*

in a remote sensing image or sampling intervals in an ecological study. In the ecological literature, resolution is often referred as "grain." In the past, because of limitation of data storage capacity, studies of large spatial extent were often associated with coarse resolution, and fine resolution was characteristic of small-scale studies. In this information age, however, where data storage becomes a lesser problem, fine-resolution data in large-scale studies are increasingly common. It is important to distinguish that once spatial data such as maps are encoded digitally, their resolutions are fixed, and that zooming in during display can only increase its cartographic scale but not its measurement scale (i.e., resolution).

4 The *operational* scale refers to the spatial extent at which certain processes operate in the environment, and may take from several to hundreds of pixels (e.g., from meters to hundreds of meters). This concept can be easily extended to both natural and social-political processes. Some researchers refer operational scale as the "scale of action" (scale at which the pattern manifests the maximum variability), and methods have been suggested to find the "scale of action." Finding the operational scale of a phenomenon is an important step, because it determines how large the spatial extent (i.e., the observational scale) and the resolution (i.e., the measurement scale) of a study should be in order to capture the major variation or characteristics of the pattern, which ultimately affects the ability of the study in revealing the underlying processes.

In defining scale, it is important to note that not only consistent terminology must be developed, but also they must be adopted widely by researchers (UCGIS, 1998). The "science of scales" will not progress until some fundamental building blocks, in this case lexical meanings of scale, have been laid out and agreed upon. While the above definitions were devised with remote sensing and GIS applications in mind, extensions or modifications of the above definitions to physical geography and human geography applications are possible and strongly encouraged.

Scale Issues in Environmental Monitoring and Change Detection

A major application of remote sensing and GIS in environmental monitoring is to detect land-use/land-cover changes (LULCC). These changes may occur globally, regionally, and locally, and they may also occur rapidly or slowly. Scale, in both spatial and temporal domains, plays an essential role in this LULC study. Because data gathering technology changes over time, time series images of different spatial resolutions and spatial extents are often collected to evaluate land-use/land-cover changes. As with most scientific investigations, the two main issues of scale in environmental monitoring and change detection are: what are the appropriate spatial resolution and spatial coverage to be used to best measure the biophysical parameters and their changes? And, would the estimates derived from using one level of scale (i.e., spatial resolution and spatial extent) be applicable to another level of scale? These two questions have important applications in change detection studies, because associated with the land-use/land-cover changes are their ecological and economic impacts on the environment, such as deforestation, carbon sequestration, decrease in clean water supplies, soil loss, and decrease in agricultural productivity. These impacts must be evaluated and estimated based on various local studies. How to extrapolate the estimates of their impacts from local into regional scales becomes a key research and policy issue in various science communities and funding agencies such as NASA and Global Change studies committees (see for example the recent funding initiatives at NASA's web site, http://www.earth.nasa.gov/research/index.html).

The determination of appropriate resolution and coverage for an environmental monitoring study is difficult. It is both a function of the characteristics of the environment under investigation and the kind of information desired. Availability of data at specific resolution level also influences the choice of data and the outcome of the study. For example, the spatial resolution required in the study of an urban phenomenon is generally quite different from that needed to study a rural phenomenon.

The effects of pixel resolution on measurement of landscape parameters from remote sensing imagery have been well documented. Benson and Mackenzie (1995) showed that landscape parameters measured from a series of aggregated images aggregated from a 30 m Landsat-TM image change drastically. Their study demonstrated that as pixel size increases, the values of some landscape parameters increase, such as percentage of water and number of lakes, whereas other parameters decrease, such as mean lake area. Their findings are typical with most scale studies as they reflect the complex nature of scale: not only the values of landscape parameters change with resolution, they also change in different directions and rates. Thus, the unpredictable nature of scale effects poses an uncertainty in reliable modeling and estimation.

Another study by Bian (1997) documented that R^2 values of a regression between two variables, biomass index and elevation, increased from 0.46 to 0.71 when the pixel resolution changed from 1 to 75 pixels. The implication of this study is that fundamental biophysical relationships change with pixel resolution, making it difficult to extrapolate the results to larger spatial extent where pixel resolution is generally coarser.

In land-cover change detection, many methods have been developed and detailed descriptions of most traditional change-detection methods can be found in standard remote sensing texts (Lillesand and Kiefer, 1999). Two basic methods, postclassification comparison and temporal image differencing or ratioing, are discussed here to illustrate the scale issues in change detection.

Given two images with different dates, the postclassification comparison method first registers and classifies each image into different types of land-cover independently. The two classified images are then compared pixel by pixel to determine which pixel changes its land-cover class between dates. The accuracy of this method obviously depends on the accuracy of each of the independent classifications. The errors that exist in each classification will propagate in the change detection process. In temporal image differencing or ratioing, either the raw differences in digital numbers between the two images or their ratios of difference are calculated. A meaningful threshold of change or no change must then be applied to determine whether the difference signals real change between the two dates. The accuracy of this approach depends upon how meaningful is the threshold, and whether the same threshold is applicable for all land-covers throughout the images. For both methods, the spatial resolution of pixels affects the accuracy in determining where the changes occur, and the temporal and radiometric resolutions of the images affect the accuracy in determining whether the changes are real. Although new methods or variations of existing methods for change detection are regularly reported in the literature, which have substantially improved the procedure of change detection and its accuracy, inherent difficulties involved in land-cover change detection remain. These difficulties arise from many factors, such as atmospheric or sun angle differences between images of different dates, pixel mis-registration, the lack of appropriate data for earlier dates, the need to integrate multiresolution and multisensor data, and the lack of assessment of the results. But most of all, the scale effects (primarily resolution and spatial extent), which have been documented above for studies involving only one image, will compound quickly in change detection studies when two or more images are used.

The Fractal Method

The fractal method is demonstrated here as a scale-based method for characterizing landscapes, landscape complexity, and their scaling properties. Fractal analysis has been suggested as a useful technique for characterizing remote sensing images as well as identifying the effects of scale changes on the properties of images (De Cola, 1993; Lam, 1990; Lam and Quattrochi,

1992; Emerson et al., 1999). Lately, Lam and others (1998; 2002) proposed the use of fractal dimension as a fundamental spatial index for characterizing spatial patterns at different ranges of scale, as part of the metadata for data mining and retrieval, and for rapid change detection.

Fractal geometry was derived by Mandelbrot (1983) mainly because of the difficulty in analyzing spatial forms and processes by classical geometry. In classical (Euclidean) geometry, a point has an integer topological dimension of zero, a line is one-dimensional, an area has two dimensions, and a volume three dimensions. In fractal geometry, the *fractal dimension* (D), is a noninteger value that exceeds the Euclidean topological dimension. As the form of a point pattern, a line, or an area feature grows more geometrically complex, the fractal dimension increases. The fractal dimension of a point pattern can be any value between zero and one, a curve between one and two, and a surface between two and three. For example, coastlines generally have dimensions typically around 1.2, and relief dimensions around 2.2. For landscape surfaces manifested in remote sensing images, fractal dimensions are generally higher, with values close to 2.7 (Qiu et al., 1999; Emerson et al., 1999).

Self-similarity is the foundation for fractal analysis (Mandelbrot, 1983), and is defined as a property of a curve or surface where each part is indistinguishable from the whole, or where the form of the curve or surface is invariant with respect to scale. The degree of self-similarity, expressed as a self-similarity ratio, is used to define the theoretical fractal dimension. In practice, the D value of a curve (e.g., a coastline) is estimated by measuring the length of the curve using various step sizes. The more irregular the curve, the greater the increase in length as step size increases. D can be calculated by the following equations:

$$\text{Log } L = K + B \text{ Log } \delta \quad [1]$$

$$D = 1 - B \quad [2]$$

where L is the length of the curve, δ is the step size, B is the slope of the regression, and K is a constant. From the above equations, D is a function of the regression slope B. The steeper the negative slope (B is a negative value), the higher the fractal dimension. The D value of a surface can be estimated in a similar fashion, and several algorithms for measuring surface dimension have been developed (Lam and De Cola, 1993).

The regression relationship between step size and line length in logarithmic terms can be illustrated visually by a scatter plot, which is commonly known as the fractal plot. An example of a fractal plot using the Louisiana coastline is shown in Figure 1.2 (Lam and Qiu, 1992). The overall fractal dimension for the entire coastline was estimated to be 1.27. When comparing different portions of the coast, the straighter portion in the west (A – Chenier plain) yielded a dimension of 1.004, whereas the complex deltaic portion in the east (B) was estimated to be 1.37. The measured fractal dimensions reflect well the underlying coastal processes, with the smoother portion of the coast a result of being dominated by the marine process, and the contorted portion dominated by the river deltaic depositional process.

The fractal plot is a useful device to illustrate the scaling properties of the phenomenon. True fractals with self-similarity at all scales (i.e., step sizes)

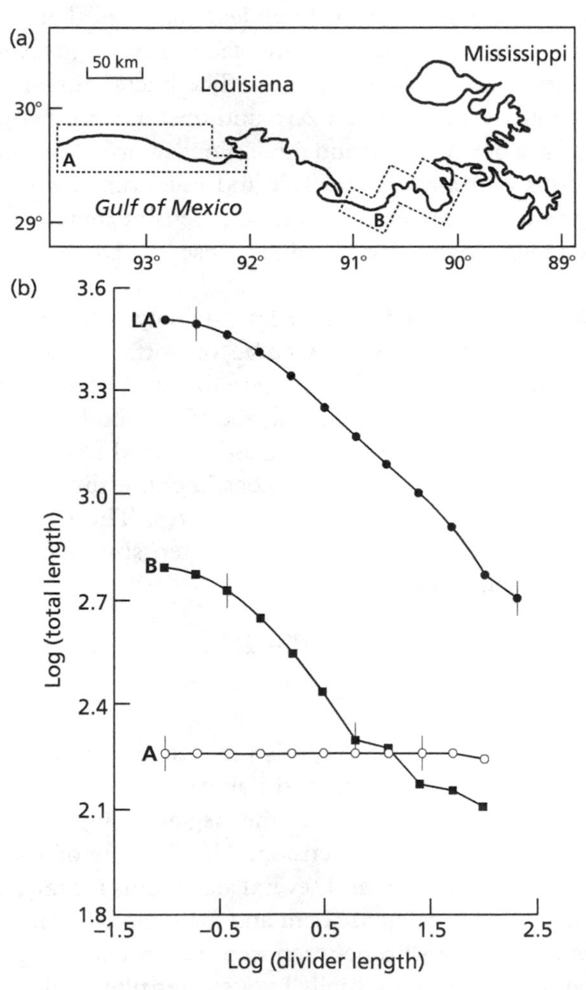

Figure 1.2 *A map (a) and fractal plot (b) of the Louisiana coastline (LA) and some of its subdivisions. A–Chenier plain and B–Plaquemines and Balize subdeltas. The bars on each curve indicate the scale ranges over which the fractal dimensions were calculated.*

will lead to a linear fractal plot. However, many previous studies have shown that fractal plots of empirical curves and surfaces are seldom linear throughout all scales. Instead, as Figure 1.2 shows, linearity occurs only within certain scale ranges, with obvious breaks between them. This indicates that true fractals with self-similarity at all scales are uncommon and unrealistic, and the notion of strict self-similarity becomes a major criticism of fractals. Recent developments in fractals have aimed at relaxing such requirement, and concepts such as statistical self-similarity, self-affinity, and multifractals have been suggested to make fractals more widely applicable (Feder, 1998; Lavallee et al., 1993)

Applications of Fractals in Scale Analysis

Determining scale ranges

As discussed above, many researchers nowadays realize that strict self-similarity is rare in natural phenomena. Rather than using D in the strict sense as originally defined by Mandelbrot (1983), self-similarity must be estimated statistically and at certain scale ranges. Lam and Quattrochi (1992) suggested that information on the changes of D with scale (step size, distance range) could be used to summarize the scale changes of the phenomena and help identify their underlying processes at specific ranges. In the case of Louisiana coastline, coasts dominated by deltaic processes generally have smaller scale ranges (0.4 to 12.8 km, see Figure 1.2), whereas coasts dominated by the marine processes have larger scale ranges (0.2 to 51.2 km).

Another example is provided here to show how fractals can be used to determine the scale ranges of different spatial patterns. Figure 1.3 is a fractal plot showing the relationship between variance and distance in logarithm form for three major cancer mortality patterns (stomach, esophageal, and liver) in the Taihu Region of China. This plot is basically a variogram plot in logarithm form. A common method for estimating the fractal dimension of surface is based on the slope of these logarithmically transformed variograms (Mark and Aronson, 1984). Fractal dimensions can be determined by the equation $D = 3 - (B/2)$, where B is the slope of the estimated regression line through the respective points in the variogram.

As expected, the linearity exhibited on the variogram plot for all three cancer patterns exists only within certain scale ranges. In this example, the fractal dimensions were estimated to be 2.86, 2.76, and 2.71 for liver, stomach, and esophageal cancers, respectively. The self-similar scale ranges for liver and esophageal cancers were about 5 to 150 km, while the range for stomach cancer was about 5 to 90 km. Lam and others (1993) speculated

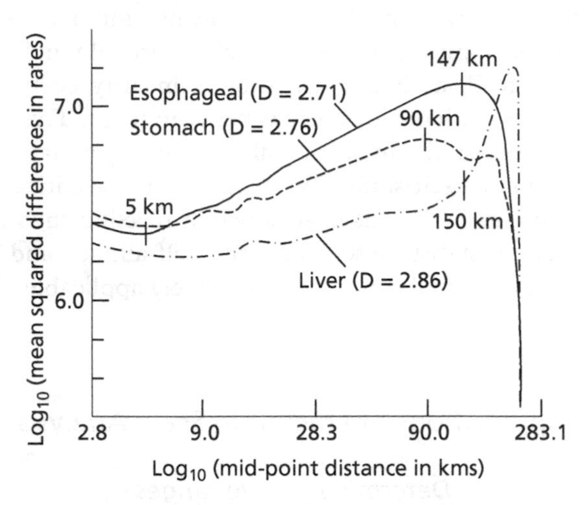

Figure 1.3 *A log-transformed variogram plot of the three major cancer mortality patterns (stomach, esophageal, and liver) for the Taihu Region, China. The bars on each curve indicate the scale ranges over which the fractal dimensions were calculated.*

that if the communes were aggregated to a distance range beyond the respective limits, the cancer patterns would look very different and would result in a different interpretation of the underlying processes. Furthermore, it was suggested that whatever the underlying controlling factors (e.g., climate, topography, pollution, economics), they are likely to operate in scale ranges similar to those of the cancer mortalities.

Effects of scale on landscape characterization

Although fractals have been suggested as a useful technique for characterizing land-covers and landscape complexity from remote sensing images for quite some time (Lam, 1990), studies on their applicability to a variety of images have been very few. One major factor contributing to the scarcity of the studies is that fractal measurement algorithms are not easily accessible to researchers. Another inhibiting factor is that different remote sensing images have different pixel resolutions and spectral bands, and their spectral response surfaces are further complicated by different sensor specifications and atmospheric effects, thus making clear and accurate land-cover characterization by fractals (or any other spatial techniques) much more difficult. Many more studies involving multisensor, multiscale, and multitemporal images are therefore needed to achieve consistent and reliable land-cover characterization.

In an attempt to overcome the first inhibiting factor, Lam and others developed a software module called ICAMS (Image Characterization And Modeling System) to provide fractal measurement and other spatial techniques for analyzing multiscale remote sensing data (Lam et al., 1998). Using the ICAMS software for fractal computation, we demonstrate in the example below how fractal dimension can be used to characterize different land-covers based on their spatial complexity, and how such complexity measures change with changes in pixel resolution.

Figure 1.4 shows four study areas (land-cover types) extracted from an ATLAS (Advanced Thermal and Land Application Sensor) image of Atlanta, Georgia (data provided by Quattrochi and Luvall 1997). The ATLAS is a 15-band multispectral scanner. A description of the ATLAS sensor and its system specifications can be found in Quattrochi and Luvall (1997). The images shown in Figure 1.4 (512×512 pixels with a pixel resolution of 10 m) are their NDVI values. NDVI (Normalized Difference Vegetation Index) values are ratios between red and infrared bands and are typically used in remote sensing to identify broad land-cover types and to serve as an important ecological indicator. In this example, NDVI were derived from:

$$\text{NDVI} = (\text{Band } 6 - \text{Band } 4)/(\text{Band } 6 + \text{Band } 4). \tag{3}$$

In general, water, moist soil, and bright nonvegetated surfaces have NDVI values less than zero and are displayed as darker shades in a gray-scale image. Rock and dry bare soils have values close to zero, and are shown in brighter shades. Positive NDVI values generally indicate green vegetation and are the brightest in the image.

The four study areas include areas from the Atlanta CBD, Hartsfield airport, mixed land-use southwest of CBD, and forested area in the north. These four land-cover types represent typical landscapes found in large metropolitan areas such as Atlanta. The triangular prism method in ICAMS was applied to compute the fractal dimension for the four NDVI images (Figure 1.5). It is clearly shown from Figure 1.5 that the airport scene has the lowest fractal dimension and the CBD scene has the highest, reflecting their spatial complexity respectively. The airport scene has a relatively uniform land-cover, with big blocks of buildings, roads, and parking areas dominating the scene. On the contrary, the CBD has a variety of land-covers, intermingling together to form a spatially complex surface of NDVI values. The mixed land-use scene and the forested scene are very similar in spatial complexity, as they consist of similar combinations of land-cover.

However, the above results can change if pixel resolution changes. To simulate the effects of changing pixel resolution on the fractal dimension

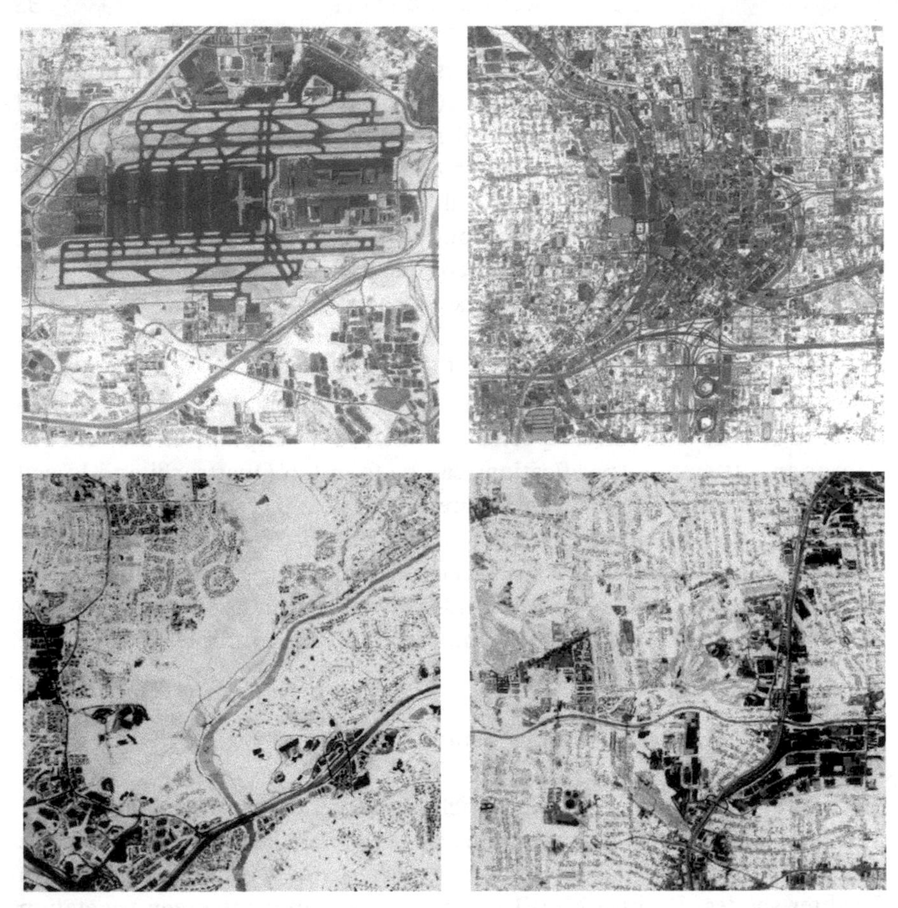

Figure 1.4 *NDVI (Normalized Difference Vegetation Index) images (512×512 pixels with a pixel resolution of 10 m) of the four study areas derived from an ATLAS image of Atlanta, Georgia. From upper left clockwise are airport (lowest fractal dimension), CBD (highest fractal dimension), forest, and mixed land-cover types.*

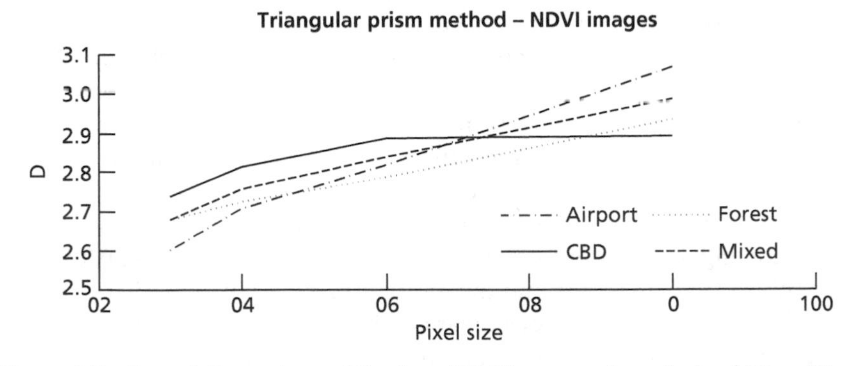

Figure 1.5 *Fractal dimensions of the four NDVI images using pixels of 10 m, 20 m, 40 m, and 80 m.*

values, the original 10 m pixels were aggregated into pixels of 20 m, 40 m, and 80 m, and their fractal dimensions were recomputed. Figure 1.5 clearly shows that as pixel size increases, fractal dimension values increase, and at approximately 50 m, all but the forest scene have roughly the same dimensions. At 80 m, however, the trend reverses, with the NDVI of the CBD scene yielding the lowest dimension and the NDVI of the airport scene returning the highest dimension. Although this kind of scale analysis has been pointed out to be possibly an artifact of the aggregation method (Weigel, 1996), the example clearly shows that the measurement scale (pixel resolution) has significant impacts on the resultant indices and hence interpretation of landscape complexity. Likewise, ecological indicators estimated from these images, such as NDVI, biomass, wetland acreages, and natural and human delineated boundaries, would also be affected. More importantly, the example reveals an intriguing pattern: that a spatially homogeneous surface (airport) is more sensitive to resolution changes than a spatially heterogeneous surface (CBD). In other words, spatially complex patterns seem to be more robust to scale effects.

Local fractals and change detection

Recently, fractals have been proposed as a useful technique for land-cover change detection (Lam et al., 1998; Emerson et al., 1999). The proposition is based on the fact that spatial indices, such as fractal dimension, are considered useful in measuring the spatial properties of landscapes from a single image. By extension, spatial indices measured from time-series images can be compared to detect changes. It is argued that this spatial approach offers some theoretical advantages over most conventional approaches for change detection (e.g., postclassification comparison). In the spatial approach, the spatial relationships among pixels, instead of the spectral values of individual pixels, are measured and compared. If there is only minor difference between the two images, the spatial relationship is not likely to change and thus the spatial indices will remain the same or similar. Conversely, only significant changes in values between the two images will lead to changes in spatial relationship and their resultant indices. In other words, measuring the spatial structure of images can better reveal the dominant pattern of the image and avoid spurious changes that might result from factors other than real land-cover change (e.g., different atmospheric conditions). This spatial approach is expected to be less impacted by the systematic or registration errors that often occur when time-series imagery is used.

Moreover, unlike many spatial indices used in landscape ecology (e.g., contagion, dominance), the fractal technique can be applied directly to images that have not been classified, thereby avoiding the tedious process of image classification that is required in most conventional change detection approaches. This property makes the fractal technique potentially a useful tool for rapid change detection and monitoring.

However, for accurate change detection, the traditional manner of applying fractal analysis to a large area to derive one single index will not work. Scale plays a key role here: fractal analysis must be applied in a hierarchical or multiscale manner, from global (large extent) to local (small extent), to accurately detect changes. We are currently working on improving the ICAMS software to incorporate a module for computing local fractals for change detection. The following example illustrates how the multiscale approach would work upon completion of the software.

Figure 1.6 shows a study area from Lake Charles, Louisiana in two images of different dates (1984 and 1993 Landsat-TM images) using their NDVI values. The study area is defined by 201 × 201 pixels with a pixel size of 25 m. Lake Charles is located in the southwestern portion of Louisiana. It had a population of about 75,000 in 1980, which decreased to 71,000 in 1992. In general, not much development or land-cover change is expected in this small metropolitan area during the 10-year period. A visual comparison of the two images (Figure 1.6) shows that the largest visible difference occurred in the lower right hand corner of the images.

The triangular prism method in ICAMS was used to compute the fractal dimensions of the two images. First, fractal dimensions for the entire images were computed, then the images were divided into four quadrants, and fractal dimensions were recomputed for each quadrant. The results

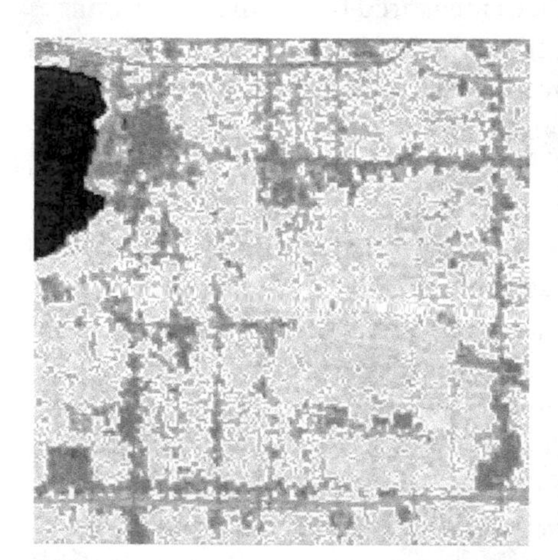
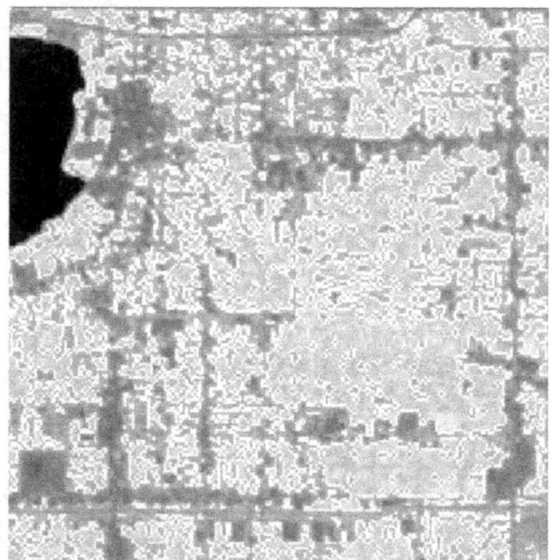

Figure 1.6 *NDVI images of 1984 (left) and 1993 (right) derived from Landsat-TM images of Lake Charles, Louisiana. The images have 201×201 pixels with a pixel resolution of 25 m.*

(Table 1.1) show that the two "global" fractal values are very similar (2.72 versus 2.68), indicating only slight changes between the two dates for the entire study area. When zoomed in with a smaller spatial extent, the largest difference in fractal dimension (2.78 versus 2.74) was found in the lower right quadrant, a result corresponding to our visual examination.

However, to pinpoint where exactly the changes are, further subdivision into smaller windows or the employment of a moving-window is necessary to more accurately identify the location of changes. As mentioned above, we are currently working on the implementation of these multiscale modules to further test the use of fractals (as well as other spatial techniques) in change detection.

Conclusions

The fractal technique is a scale-based technique. By this, we mean that the theoretical concept used to define fractals as well as the practical algorithms used to measure the fractal dimension are affected by the four elements of scale. The original fractal model was based on the self-similarity property, which assumes that the form or pattern of the spatial phenomenon remains unchanged throughout all scales. Rather than using the fractal dimension in the strictest sense, this chapter illustrates that it is possible to use the fractal dimension as a parameter to summarize the scale changes and their effects. Through interpretation of how the fractal dimension values change with scale (i.e., the step size used to compute the dimension), it may be possible to relate or separate scales of variation that might be the result of particular spatial processes. The three examples used in this chapter demonstrate that the fractal technique can be used to determine the scale ranges of spatial phenomena; examine the relationship between spatial complexity and the effects of scale on spatial complexity; and detect land-cover changes through a multiscale approach.

The main goal of this research is to identify effective indices and measurement techniques for rapid change detection and environmental monitoring and to develop techniques to mitigate the scale effects.

Table 1.1 *Fractal dimensions of the NDVI (Normalized Difference Vegetation Index) images from Lake Charles, Louisiana. Upper left (UL), upper right (UR), lower left (LL), and lower right (LR) quadrant*

	Whole	*UL*	*UR*	*LL*	*LR*
1984	2.72	2.69	2.77	2.74	2.78
1993	2.68	2.67	2.75	2.77	2.74

Ultimately, these indices could be used to reflect the underlying landscape patterns and processes, identify the "characteristic" or optimal scale of analysis, and form a part of the metadata of the image that could then be used as a guide to search an image for spatial data mining or rapid change detection. Because different spatial phenomena have different scaling properties, it is important to have scale-based methods such as fractals to identify and measure such properties so that uncertainties due to scale can be reduced in subsequent analyses.

The application of fractals in change detection is rather new, and many more experiments are needed to identify the various technical problems associated with this type of application as well as to suggest possible improvements of the approach. As shown in this chapter, the fractal technique can be applied in a multiscale manner, from global to local, to achieve higher accuracy in change detection. This is similar to Hou's study (1998), which demonstrated with a striking example the effectiveness of local detection in detecting changes in forest composition. Likewise, the multiscale wavelet technique, which captures the characteristics of landscape by decomposing it into different resolution levels, has been shown in two recent studies to be able to classify some land-covers with 100 percent accuracy (Zhao, 2001; Myint, 2001). These scale-based methods offer a promising avenue for change detection and environmental monitoring.

Acknowledgments

This research is partially supported by a research grant from NASA (award number: NAGW-4221) and by a NASA Summer Faculty Fellowship. Additional support is provided through a NASA Intelligent Systems research grant (NAS-2-37143).

References

Atkinson, P. M., and N. J. Tate (eds) 1999: *Advances in Remote Sensing and GIS Analysis*. London: John Wiley.

Benson, B. J., and M. D. MacKenzie 1995: Effects of sensor spatial resolution on landscape structure parameters. *Landscape Ecology* 10(2): 113–20.

Bian, L. 1997: Multiscale nature of spatial data in scaling up environmental models. In D. A. Quattrochi and M. F. Goodchild (eds), *Scale in Remote Sensing and GIS*. Boca Raton, FL: CRC/Lewis Publishers, 3–26.

Cao, C., and N. S.-N. Lam 1997: Understanding the scale and resolution effects in remote sensing and GIS. In D. A. Quattrochi and M F. Goodchild (eds), *Scale in Remote Sensing and GIS*. Boca Raton, FL: CRC/Lewis Publishers, 57–72.

De Cola, L. 1993: Multifractals in image processing and process imaging. In N. S.-N. Lam and L. De Cola (eds), *Fractals in Geography*. Englewood Cliffs, NJ: Prentice Hall, 282–304.

Emerson, C. W., N. S.-N. Lam, and D. A. Quattrochi 1999: Multiscale fractal analysis of image texture and pattern. *Photogrammetric Engineeing and Remote Sensing*, 65(1): 51–61.

Goodchild, M. F. 1986. *Spatial Autocorrelation*. CATMOG (Concepts and Techniques in Modern Geography) 47. Norwich, UK: Geo Books.

Feder, J. 1988: *Fractals*. New York: Plenum Press.

Harvey, D. 1968: Processes, patterns and scale problems in geographical research. *Transactions of the Institute of British Geographers* 45: 71–8.

Hou, R.-R., 1998: *A local-level approach in detecting scale effects on landscape indices*. Master Thesis, Department of Geography and Anthropology, Louisiana State University.

Lam, N. S.-N. 1990: Description and measurement of Landsat TM images using fractals. *Photogrammetric Engineering and Remote Sensing* 56(2): 187–95.

Lam, N. S.-N., and H.-L. Qiu 1992: The fractal nature of the Louisiana coastline. In D. Janelle (ed.), *Geographical Snapshots of North America*. New York: The Guilford Press, 270–4.

Lam, N. S.-N., and D. A. Quattrochi 1992: On the issues of scale, resolution, and fractal analysis in the mapping sciences. *The Professional Geographer* 44(1): 89–99.

Lam, N. S.-N., and L. De Cola (eds) 1993: *Fractals in Geography*. Englewood Cliffs, NJ: Prentice Hall, 308.

Lam, N. S.-N., H. L. Qiu, R. Zhao, and N. Jiang 1993: A fractal analysis of cancer mortality patterns in China. In N. S.-N. Lam and L. De Cola (eds), *Fractals in Geography*, Englewood Cliffs, NJ: Prentice Hall, 247–61.

Lam, N. S.-N., D. A. Quattrochi, H.-L. Qiu, and W. Zhao 1998: Environmental assessment and monitoring with image characterization and modeling system using multiscale remote sensing data. *Applied Geography Studies* 2(2): 77–93.

Lam, N. S.-N., H.-L. Qiu, D. A. Quattrochi, and C. W. Emerson 2002: An evaluation of fractal methods for characterizing image complexity. *Cartography and Geographic Information Science* 29(1): 25–35.

Lavallee, D., S. Lovejoy, D. Schertzer, and P. Ladoy 1993: Nonlinear variability of landscape topography: Multifractal analysis and simulation. In N. S.-N. Lam and L. De Cola (eds), *Fractals in Geography*, Englewood Cliffs, NJ: Prentice Hall, 158–92.

Lillesand, T. M., and R. W. Kiefer 1999: *Remote Sensing and Image Interpretation*. 4th edn, New York: Wiley.

Mandelbrot, B. B. 1983: *The Fractal Geometry of Nature*. New York: W. H. Freeman.

Mark, D. M., and P. B. Aronson 1984: Scale-dependent fractal dimensions of topographic surfaces: An empirical investigation, with applications in geomorphology and computer mapping. *Mathematical Geology* 11: 671–84.

Moellering, H., and W. R. Tobler 1972: Geographical variances. *Geographical Analysis* 4: 35–50.

Myers, D. E. 1997: Statistical models for multiple-scaled analysis. In *Advances in Remote Sensing and GIS Analysis*, P. M. Atkinson and N. J. Tate (eds), London: John Wiley & Sons Ltd., 273–93.

Myint, S. W. 2001: *Wavelet Analysis and Classification of Urban Environment Using High-Resolution Multispectral Image Data*. Doctoral Dissertation, Department of Geography and Anthropology, Louisiana State University.

Nellis, M. D., and J. M. Briggs 1989: The effect of spatial scale on Konza landscape classification using textural analysis. *Landscape Ecology* 2: 93–100.

O'Neill, R.V., A. R. Johnson, and A. W. King 1989: A hierarchical framework for the analysis of scale. *Landscape Ecology* 7(1): 55–61.

Openshaw, S. 1984: *The Modifiable Areal Unit Problem*. CATMOG (Concepts and Techniques in Modern Geography) Series 38. Norwich, UK: Geo Abstracts.

Qiu, H.-L., N. S.-N. Lam, and D. A. Quattrochi 1999: Fractal characterization of hyperspectral imagery. *Photogrammetric Engineering and Remote Sensing* 65(1): 63–72.

Quattrochi, D. A., and M. F. Goodchild 1997: *Scale in Remote Sensing and GIS*. Boca Raton, FL: CRC Press.

Quattrochi, D.A., and J. Luvall 1997: High spatial resolution airborne multispectral thermal infrared data to support analysis and modeling tasks in EOS IDS Project ATLANTA. *The Earth Observer* 9(3): 22–7.

Stone, K. H. 1972: A geographer's strength: the multiple-scale approach. *Journal of Geography* 71(6): 354–62.

Tate, N., and P. M. Atkinson (eds) 2001: *Modeling Scale in Geographical Information Science*. Chichester, UK: John Wiley.

Tobler, W. R. 1988: Resolution, resampling, and all that. In *Building Database for Global Science*, H. Mounsey and R. Tomlinson (eds), London: Talyor & Francis, 129–37.

Turner, M. G., V. H. Dale, and R. H. Gardner 1989: Predicting across scales: theory development and testing. *Landscape Ecology* 3(3/4): 245–52.

University Consortium for Geographic Information Science 1998: Research Priorities. http://www.ucgis.org/research98.html.

Watson, M. K. 1978: The scale problem in human geography. *Geografiska Annaler* 67: 83–8.

Weigel, S. J. 1996: *Scale, Resolution and Resampling: Representation and Analysis of Remotely Sensed Landscapes Across Scale in Geographic Information Systems*. Doctoral Dissertation, Department of Geography and Anthropology, Louisiana State University.

Woodcock, C. E., and A. H. Strahler 1987: The factor of scale in remote sensing. *Remote Sensing of Environment* 21: 311–32.

Zhao, W. 2001: *Multiscale Analysis For Characterization of Remotely Sensed Images*. Doctoral Dissertation, Department of Geography and Anthropology, Louisiana State University.

Zhu, C., and X. Yang 1998: Study of remote sensing image texture analysis and classification using wavelet. *International Journal of Remote Sensing* 13: 3167–87.

2 Population and Environment Interactions: Spatial Considerations in Landscape Characterization and Modeling

Stephen J. Walsh, Kelley A. Crews-Meyer, Thomas W. Crawford, William F. Welsh

Geographers and others have recently begun applying the scale–pattern–process paradigm to the study of population–environment interactions. This paradigm holds that the scale at which the landscape is observed affects discernible patterns that may be related to processes at work on the landscape. This emergent area of study emphasizes the interplay between social, biophysical, and geographical factors as drivers of landscape change and suggests the relevance of scale in the dependence of their interrelationships.

The land-use/land-cover change (LULCC) community has been active in examining population-environment interactions and the spatial pattern of their landscape representations through meetings (e.g., Association of American Geographers 1995–2001, Population Association of America 1995–2001, Open Meeting of the Human Dimensions of Global Environmental Change Research Community – Vienna 1997, Barcelona 1998, Tokyo 1999, Rio de Janeiro 2001) and publications such as *People and Pixels: Linking Remote Sensing and Social Science* (Liverman et al., 1998), *Proceedings of PECORA 13 Symposium on Human Interactions with the Environment: Perspectives from Space* (American Society for Photogrammetry and Remote Sensing, 1998), *Remote Sensing and GIS Applications in Biogeography and Ecology* (Millington et al., 2000), and *Remote Sensing and GIS Applications for Linking People, Place, and Policy* (Walsh and Crews-Meyer, 2001). Consistent among these recent conferences, workshops, and publications is an emphasis on defining drivers of LULCC, and the realization that multithematic and multiscale perspectives are essential to effectively address complex human–environment interactions and their linkage to landscape form and function. Quattrochi and Goodchild (1997) in their edited volume, *Scale in Remote Sensing and GIS*, presented the results of research efforts that examined scale and used Geographic Information Science (GISc) methods to characterize landscape patterns and processes primarily within the natural sciences. Here, social and natural sciences are integrated within a spatial analytical context and through a scale-dependent framework.

Concepts from landscape ecology have been used to substantiate the multiscale perspective and to provide context to the study of scale-pattern-process as it relates to population and the environment. Landscape ecology holds that all landscapes contain structure, function, and change (Forman and Gordon, 1986). *Structure* refers to the spatial relationships among the distinctive elements or ecosystems comprising a landscape, with the sizes, shapes, and configurations of these elements being of key importance. *Function* refers to spatial and temporal flows of energy, material, and species among landscape components, and *change* refers to the alteration in structure and function of the ecological mosaic over time. Rooted in landscape ecology is the paradigm that the spatial pattern of a landscape strongly influences the ecological characteristics of that landscape, and that the relationships among ecosystems are interactive, dynamic, and scale dependent (Risser et al., 1984; Urban et al., 1987; Turner, 1990a; Turner, 1990b; Walsh et al., 1994b).

The scale dependence of ecological relationships is underscored by hierarchy theory which posits that scale is an epistemological construct and that "scale-dependent" entities do not *exist* at certain scales; rather, they require multiple scales of perception to make them appear in certain ways. A corollary to this position is that multiscale observation is necessary for a more complete understanding of the variety of landscape patterns and processes. More specifically, researchers should take the scale of inquiry for any phenomenon and observe the scale "above" (over a larger time or spatial extent) for context and the scale "below" (lesser spatial or temporal extent) for the content or mechanics of the phenomenon of interest (Crews-Meyer, 1999, 2000). Geographers are well positioned to implement principles of hierarchy theory and scale in their research, especially through the use of GISc methodologies.

GISc approaches are being used here to study population and the environment. land-use/land-cover (LULC), change trajectories of LULC, metrics of landscape organization, and plant biomass (as used here) are often cast as the dependent variable in population-environment research, but demographic or behavioral variables also serve as dependent variables in multivariate analyses. Independent variables might include climatic data from weather stations, measures of social networks that connect villages or districts through institutional or commodity sharing, ecological disturbances and patterns, commodity prices, terrain descriptors, and demographic characteristics of households or villages. Such variables can be characterized over time through satellite systems such as Landsat that provide historical snapshots of the landscape beginning in 1972 and extending through the present, and aerial photography that offers landscape views often times over a half century or more. In addition to being time-sensitive (retrospective or prospective views are possible), GISc can be space-sensitive in that data can be captured at a variety of grains and extents to accommodate scale dependent studies. In addition, spatial analytical techniques can be used (e.g., agglomeration techniques, social or environmental hierarchies) for the iterative recomputing of cell sizes used to partition space and summarize landscape features. Finally, GISc is capable of creating numerous types of outputs to accommodate the diversity of data, researchers, and problems being addressed. The visualization of derived or value-added data is a common approach for examining graphically the spatial and/or temporal dynamics of parameters and systems.

Research Objectives

The basic intent of this research is to examine scale dependent relationships between plant biomass (green canopy vegetation) and a selected set of social and biophysical variables for a region in rural northeast Thailand. Multivariate models are developed at nine scale steps extending from 30 to 1050 m cells to examine scale-pattern-process relationships between terrain characteristics, representing site conditions and resource potential, and population variables, representing demographic characteristics of villages and the ability of villages to alter the landscape. The study integrates terrain data derived from digital elevation models and population data collected during a 1994 social survey of all 310 villages within the study area. Plant biomass, the dependent variable, is estimated by applying the corrected version of the Normalized Difference Vegetation Index (NDVI) to a Landsat Thematic Mapper (TM) image. This research does not intend to either describe a model of LULCC or to exhaustively explore the population and environment implications of observed scale dependencies. Rather, it offers (i) an application of techniques used to explore the scale dependence of relationships between plant biomass variation and biophysical and social variables, (ii) a description of data manipulation considerations for merging disparate data within a spatially-explicit context, and (iii) a preliminary interpretation of the relationships between plant biomass and social/environmental variables occurring in an agrarian and former frontier setting across a range of spatial scales.

Literature Review and Theoretical Justification

Landscape ecology, human ecology, and political ecology serve as the theoretical context for this research. Landscape ecology examines the interrelationship of scale–pattern–process; human ecology states that people are important actors on the landscape that shape and are shaped by their physical and social environments; and political ecology emphasizes the influence of broad-scale political and economic structures on the social-biophysical environment.

Vink (1983) indicates that landscape ecology has two components: it is a research approach that considers landscapes to support both natural and cultural ecosystems; and it is a "science that investigates the relationships between the biosphere and anthroposphere and either the Earth's surface or the abiotic components" (Goudie et al., 1994: 301). The determination of the distinctive elements or ecosystems comprising any given landscape – and thus the definition of the landscape itself – is a function of scale, making the study and characterization of spatial and temporal relationships a major focus in landscape ecology research (e.g., Forman and Godron, 1986; Meentemeyer and Box, 1987; Bian and Walsh, 1993; Cousins, 1993; Klijn and de Haes, 1994; Luque et al., 1994; Walsh et al., 1994a, b; Benson and MacKenzie, 1995; Moody and Woodcock, 1995; Qi and Wu, 1996; Lambin, 1997; Walsh et al., 1997). The inclusion of human influences with biophysical factors within a landscape further makes landscape ecology an appropriate perspective from which to examine scaling relationships between population and environment interactions (e.g., Vink, 1983; Forman, 1995; Nassauer, 1995). However, landscape ecology tends to focus on the material manifestation of landscapes (i.e., flora, fauna, the spatial imprint of land-use practices, and the built environment) and does not purport to be theoretically and methodologically prepared to offer explanations of social processes and patterns. However, the methods and theories developed in human and political ecology can provide a means to characterize and address the social implications of scale exhibited through landscape patterns.

Human ecology is an extension of the concept of biological ecology, which studies the relationships between organisms and their biophysical and social surroundings (Johnston et al., 1994). Barrows (1923) contends that geography as human ecology should be nomothetic and focus on the relationships between natural environments and the distributions and activities of humans, with emphasis on human adjustments to nature. Both sociological and geographical human ecology have been influential paradigms in their respective disciplines, and as such have been critiqued widely. What has emerged in recent decades is the important subfield of cultural ecology, with research most often focusing "on human adjustments to the natural environment, emphasizing the interactive and adaptive character of the human–environment interaction, . . . and its mediation by social institutions" (Johnston et al., 1994: 258).

The Thailand study area has been traditionally dominated by peasant subsistence agriculturalists, who in recent times have experienced significant social and environmental changes associated with deforestation, agricultural extensification, declining human fertility, significant out-migration, integration into the market economy, and increasing off-farm employment. The theory of political ecology provides a means to account for such scale-related interactions. Blaikie and Brookfield (1987: 14) state that "the scale issue is crucial to the definition of land management because it focuses on the boundary problem of decision-making and of allocating costs and benefits." They identify the importance of interactive and feedback mechanisms, geographical scales and hierarchies of socioeconomic organizations, and contradictions between environmental and social change over time as being defining characteristics in the study of complex relationships inherent to land degradation and society. The approach taken by Blaikie and Brookfield is termed *regional political ecology*. *Regional* is believed to be essential "because it is important to take account of environmental variability and the spatial variations in resilience and sensitivity of the land, as different demands are put on the land through time," and also because *regional* "implies the incorporation of environmental considerations into theories of regional growth and decline" (Blaikie and Brookfield, 1987: 17).

Study Area: Nang Rong District, Northeast Thailand

Nang Rong district is part of Buriram province, located in the southwest portion of the Khorat Plateau in northeast Thailand, and referred to as "Isaan" (Figure 2.1). Isaan contains approximately one-third of the country's area ($170,000\,km^2$) and has a population of more than 18 million people. The dominant occupation in the region is farming and the majority of farm households own an average of three hectares of land (Ghassemi et al., 1995). Per capita income is the lowest in the country, largely because of low and unstable agricultural production resulting from erratic monsoonal rainfall and generally poor soils (Arbhabhirama et al., 1988; Parnwell, 1992; Ghassemi et al., 1995). Approximately one-third of Isaan's land

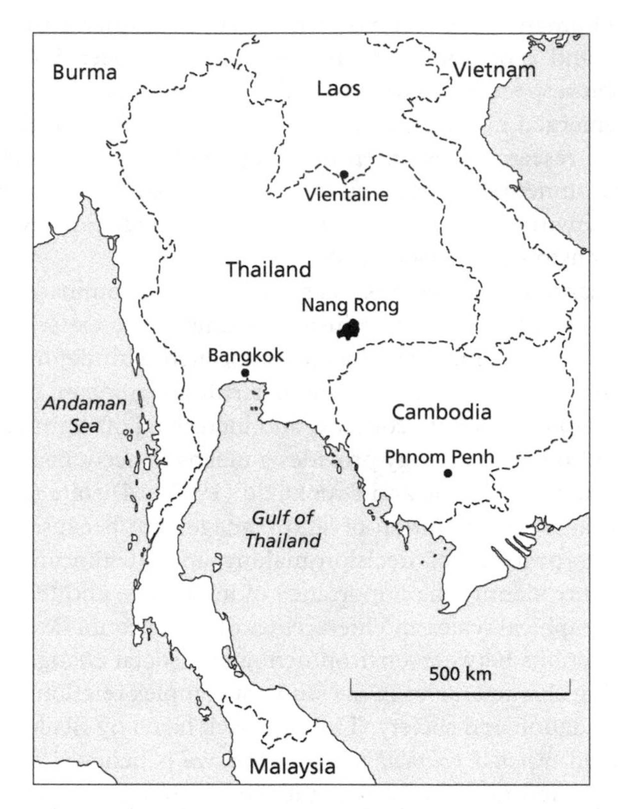

Figure 2.1 *Study area location, Nang Rong district, Northeast Thailand.*

area is unsuitable for successful cropping due either to steepness or laterization (Parnwell, 1988), and another one-third of the land is suitable for rice cultivation, though yields are relatively moderate. In many places the soils are very infertile, including areas with high levels of salinity or acidity (Ghassemi et al., 1995). Soils found around the margins of hills and on uplands are often highly susceptible to erosion.

Rainfed rice is the most important crop within Nang Rong. Paddy land occupies even marginally suitable sites. Approximately 85 percent of 7,300 sample households grow rice, according to the 1994 social survey. Upland field crops now cover large areas, mostly in the form of the cash crops, cassava and sugar cane. Perennial agriculture, including para rubber and eucalyptus groves, and tree fruit orchards are also found on selected upland sites. With a pronounced reliance on monsoonal rains, Nang Rong district is generally limited to a single rice crop each year. Irrigation in Nang Rong is limited, with 80 percent of rice growing households relying *solely* on rainfed methods.

The natural (noncultivated) vegetation of Isaan consists of a dry monsoon forest predominated by dwarf dipterocarp trees, and containing areas of grassland, thorny shrubs, and bamboo thickets (Parnwell, 1988). The vegetation phenology is largely drought-controlled (Ghassemi et al., 1995; Rundel and Boonpragop, 1995). Deforestation has been underway in the Northeast for more than a century, but the pace has been accelerated since World War II and the spatial imprint of deforestation has substantially increased. Beginning in the 1960s, and significantly increasing during the 1970s, deforestation accelerated to accommodate extensification associated with a transition from subsistence to a mixed subsistence/commercial agriculture orientation. This transition likely was related to the growth of a monetary economy, market integration, and a desire for higher living standards commonly associated with consumer goods.

Upland sites were deforested primarily for the production of field crops such as kenaf, corn, sugarcane, and especially cassava (exported to Europe for animal feed). Cassava is a drought-resistant tuber crop that is grown in marginal, upland sites in relatively large fields that coalesce into extensive regions when individual household fields are juxtaposed. Deforestation due to agricultural extensification is primarily a phenomenon seen in upper terraces and upland sites during the more recent periods. Such sites were historically viewed as marginal, or at least transitional, dependent upon innate resource limitations at localized sites, spatial and temporal vagaries of the monsoonal rains, and relatively low transportation and hydrographic accessibility. The combination of road development and land scarcity has likely influenced the spread of cultivation into these marginal sites.

Tree clearing continues in rice producing areas, and along stream courses in the alluvial plains and lower terraces, to expand and enhance rice cultivation through intensification (Fukui, 1993). As a consequence, contiguous forests have been almost entirely removed from the rice producing areas, and forests in upland sites have been substantially modified, leaving forest remnants primarily in very hilly and remote areas that are generally unsuitable for cultivation.

Data Sets and Methods

A February 1993 Landsat Thematic Mapper (TM) digital data set (scene ID# 5325902532400) was acquired to characterize the landscape at a 30×30 m pixel resolution through a level-1 LULC classification and a measure of plant biomass. The LULC was defined through an unsupervised classification of the spectral responses using the ISODATA (Interactive Self-Organizing Data Analysis Technique) clustering approach and the maximum likelihood decision-rule. Plant biomass levels were assessed through the application of a vegetation index that combines spectral channels of remotely sensed data to yield empirically derived estimates of plant biomass at the pixel level.

The mapped level-one categories included forest, water, rice, upland agriculture, and bare soil. Panchromatic aerial photographs (nominal scale of 1:50,000 for 1994) were used for class merging, labeling, and general validation. Plant biomass levels for the study area were estimated from the preprocessed (i.e., geometric and atmospheric corrections) Landsat TM data through use of the corrected version of the Normalized Difference Vegetation Index (NDVI), defined as:

$$\text{NDVI} = (\text{NIR} - \text{RED}) / (\text{NIR} + \text{RED}) \times [1\text{-}(\text{MIR} - \text{MIRmin}) / (\text{MIRmax} - \text{MIRmin})] \quad [1]$$

NIR = near-infrared spectral region; Landsat TM channel 4
RED = red-visible spectral region; Landsat TM channel 3
MIR = middle-infrared spectral region; Landsat TM channel 5

The satellite data were acquired in February when the landscape is heavily influenced by bare soil and crop stubble from the December rice harvest that is generally associated with more lowland sites. In addition, harvested and/or planted upland crops (principally cassava and sugar cane) further introduce bare soil responses to the image. The application of the index generates a continuous surface of plant biomass values, calculated initially at the 30 × 30 m spatial resolution, but aggregated through the nine selected scale steps using the base resolution for all subsequent calculations.

Three terrain variables, elevation ("elev"), slope angle ("slope"), and soil moisture potential ("wet"), were also generated through the application of GIS procedures. Digital elevation models of the study area were produced from 1984, 1:50,000 scale (contour interval of ten m) topographic basemaps for which all contour lines and all point elevations were digitized. From the derived elevation matrix, slope angle values were derived at the 30 × 30 m base resolution for subsequent spatial aggregations to the other selected scales. Elevation ("wet") was used to differentiate low, middle, and high terraces as well as the upland sites within the study area. Slope angle ("slope") was used to characterize the elevational gradient on the landscape and the topographic transition from lowland to upland sites, and soil moisture potential ("wet") was generated from the 30 × 30 m DEM using a digitized hydrography layer for hydrologic enforcement in flow routing (Walsh et al., 1998). The variable was estimated following an index of saturation potential developed by Beven and Kirkby (1979) which uses flow directions and accumulation grids based on eight directions for pixel-to-pixel flow routing after Moore et al., (1991). To reduce multicollinearity in the analyses, elevation and slope angle values were transformed by applying a multiplier (1/square root) to each variable.

The population variables used in this analysis were extracted from a 1994 survey of all 310 villages comprising Nang Rong district. The data were collected through a group interview of the village headman and other local officials. The variables included: total land under cultivation ("land"); ratio of total land under cultivation to total population ("lapo"); total population ("pop"); ratio of men to women ("sexr"); and number of households ("hous"). To reduce multicollinearity in the analyses, the ratio of total land under cultivation and total population were again transformed by 1/square root of the variable. These selected variables were extracted from an extensive set of variables to explore preliminary effects of population size, nature of the resource base, population density relative to the resource base, gender and its relationship to job sorting, and the number of household units within the village (the primary decision-making unit) that were hypothesized to affect land-use/land-cover patterns and plant biomass levels.

Integration of Social and Biophysical Data

The typical Nang Rong village displays a nuclear structure with a concentrated core surrounded by an array of agricultural fields. With land title unclear and political boundaries relatively unimportant within this context, approaches for estimating village territories have been examined through distance, geometry, and region growing approaches (Entwisle et al., 1998; Evans, 1998; Walsh et al., 1998; Crawford, 1999; 2000). Here, a 5-km radial buffer was generated around estimated village centroids to define boundaries for each of the 310 villages, and to allow the assignment of demographic variables positioned at the village-level to nearly 98 percent of the territory within the district. The buffer dimension was set through knowledge of land ownership patterns and modes of travel within the district as of 1994. Sensitivity analyses were conducted to confirm the relevancy of the 5-km spatial buffer.

There are a number of approaches to estimate village territories to address specific research objectives. For example, elliptical geometries, population weighted buffers, Thiessen polygons, and access (transportation sensitive) biased boundary dimensions are among those approaches that might be applied. This study, however, was concerned primarily with the areal association of land and people to specific villages through defined territories. Land within the district may be owned by people living outside the district, and thus issues related to social networks, kinship ties, and labor relationships may further alter the simple geometry used here to set village territories. Because households within a village are organized in a nuclear pattern, the survey data were geocoded at the estimated center-point of each village location. To be cartographically compatible with the continuous nature of the other variables generated for this analysis, the population variables were transformed from discrete to continuous by applying a population distribution model. Using the LULC classification for 1993, the landscape was stratified into agriculture (i.e., rice and upland agriculture, primarily cassava and sugar cane) and nonagriculture categories. In agricultural areas, the selected population variables were distributed in defined village territories on a per pixel basis using an equal value spread function. For territories assigned to multiple villages using the radial buffer approach, grid cells had higher variable totals as a consequence of the equal value function being implemented on a village by village basis.

Data Aggregation and Scale Levels

Through a spatial aggregation procedure, plant biomass, terrain variables, social variables, and LULC were scaled across nine spatial scales – 30, 90, 150, 210, 300, 450, 600, 900, and 1050 m cells. The scale steps represent the spatial resolution of popular sensor systems – the Landsat TM sensor at the finest spatial scale (30×30 m) and the NOAA Advanced Very High Resolution Radiometer (AVHRR) at the coarsest spatial scale ($1,100 \times 1,100$ m).

Aggregations across the nine scale steps were generated by returning to the 30×30 m base surface to initiate all computations; aggregations were not made of previously aggregated data. For the LULC scaling, a grid-based plurality procedure was used where the most frequently occurring subgrid cover type was used to code each grid cell at degraded spatial resolutions (Moody and Woodcock, 1995). Pixels of NDVI were aggregated to generate blocks of a coarser resolution image with their own degrees of variability and spatial pattern. The NDVI data were calculated from corrected spectral radiance values for each aggregation level through a nonweighted, flat-field aggregation. Each portion of the image that was defined by the boundaries of an aggregated pixel was treated as a separate block. Elevation data were aggregated in the same manner, having the effect of reducing the variability of the elevation values and biasing slope angle estimates downward.

Given the landscape characteristics of the study area, scales extending from 30–150 m represent a set of fine spatial scales hypothesized to reflect social processes influencing LULC patterns (e.g., household parcels). Scales extending from 600–1,050 m represent a set of coarse spatial scales hypothesized to reflect biophysical processes influencing LULC patterns (e.g., environmental gradients), while scales extending from 210–450 m represent a set of intermediate spatial scales hypothesized to reflect transitional processes from the social to the biophysical (e.g., household access to water). For example, individual land clearing activities instituted at the household-level to expand upland crops at the expense of forests tend to form a locally fragmented landscape matrix that extends across spatial scales via the spatial coalescence of clearings instituted by multiple households from multiple villages. The areal extension of local land clearing practiced at the fine scale but extended to coarser scales through juxtaposition yields a landscape characterized by local diversity and regional homogeneity.

A February 1, 1993 TM image was used for this research, and understanding the timing of that image is critical in characterizing plant biomass, the dependent variable. From a remote sensing perspective, the rainy period, roughly June through November, substantially affects the availability of satellite data when using optical systems such as Landsat. Cloud coverage associated with the monsoonal rains limits useable satellite scenes and constrains landscape views to the early planting period of rice cultivation and to the later harvest period in December. Other landscape views during the crop cycle are quite fortuitous, because of the clouds. In February, the rice has been harvested and stubble remains in the field. Depending upon the nearness to water, harvested rice paddies may contain grasses of varying types and densities. February images reveal a generally dry landscape, visually brown in color and stratified by topographic position. In the alluvial plains and low terraces rice is extensively cultivated, but harvested by the February image date. Isolated trees, small clusters of trees, and trees in riparian habitats are scattered throughout the rice paddies, adding greenness and plant biomass within a portion of the landscape that cycles between a chlorophyll-rich landscape and a senescent one (Kaida and Surarerks, 1984). By February, agricultural activities have become focused on the upland sites and high and middle terraces where cassava and sugar cane are predominately grown. The cassava and sugar cane fields may be vegetated or completely barren and bare, depending upon the harvest date. The native forests that remain are only remnants of past extents and densities, existing as patches in a matrix of upland field crops. In some regions of the study area, particularly in the southwest, deforestation has occurred on such a broad scale that forests are really subsumed and cassava dominates the landscape at that elevation strata. Therefore, the February landscape in the upland sites contains broad areas of secondary forest and/or field crops depending upon the location within the study area and the scale of observation.

Cassava fields are approximately 100 × 100 m, but individual fields may coalescence into broader areas as a consequence of common deforestation and agricultural extensification decisions practiced by individual households. This results in the generation of a fragmented landscape – LULC varies by the nearness to villages, site suitability's, resource endowments, and geographic accessibility to roads and water for agriculture. Plant biomass varies spatially and temporally in rhythm with the monsoonal rains, crop phenologies, and the general senescence cycle of the landscape. The rice areas are negatively correlated with plant biomass in February, and NDVI values usually increase with elevation, because senescent rice paddies are at lower elevations and forests and upland agriculture are primarily at the higher elevations. At the fine scale, forests are maintained around the nuclear village compounds, generally 100 households or smaller. These forests introduce elevated plant biomass values in rice growing, lowland areas and, together with the isolated trees that remain in rice paddies and within the riparian zones, contribute to NDVI values above those representative of a stubbled landscape. But the predominance of rice relative to the remaining trees is substantial and therefore the influence of trees at moderate to coarse scales is negligible.

Multiple Regression Analysis

Multiple regression analysis was used to examine the relationships between plant greenness and selected social and biophysical variables across the sampled spatial scales, following data processing and transformations for data compatibility. The objective was to examine the relationships between the above-described set of population and terrain variables as descriptors of plant biomass across nine scale steps ranging from 30–1,050 m. The bias towards male-dominated use of mechanization for cassava cultivation is represented in the importance of the sex ratio ("sexr") variable in villages of such predominance. Total population ("pop") reflects the ability of villages to alter the natural landscape through deforestation and to expand agriculture through extensification in the upland sites and intensification in the lowland rice-producing environments. Total land under cultivation ("land") represents the agricultural resource-base occurring within village territories, the ratio of total land under cultivation and total population reflects a measure of population density, and the number of households ("hous") indicates the number of decision-making units that are capable of altering the composition and spatial organization of land-use/land-cover types through deforestation and agricultural extensification.

Table 2.1 summarizes the regression statistics for plant biomass ("NDVI") against the set of descriptor variables: elevation ("elev"), slope angle ("slope"), soil moisture potential ("wet"), total land under cultivation ("land"), ratio of total land under cultivation and total village population ("lapo"), total population ("pop"), sex ratio ("sexr"), total number of households ("hous"). Table 2.1 also indicates the sample size of cells used to compute the regressions, slope coefficients for each variable in the model, standard error, t-values, p-values, and the adjusted R^2 for each of the nine scale steps. Note that the sample size changes with scale, affecting standard errors, t-values, and p-values. These statistics are still presented, but should be interpreted as general indicators. Also, recall that variable transformations were computed to reduce multicollinearity in the models. A random sample of agricultural cells was instituted at the fine grain scales to reduce computational intensity of the regression runs by capping the sample size below 12,000 cells or approximately 5 percent of the total available cells. Figures 2.2a and 2.2b show plots of the slope coefficients for each of the descriptor variables used in the regression models at each of the nine scale steps, and Figure 2.3 shows the variation in the adjusted R^2-values across the sampled scales.

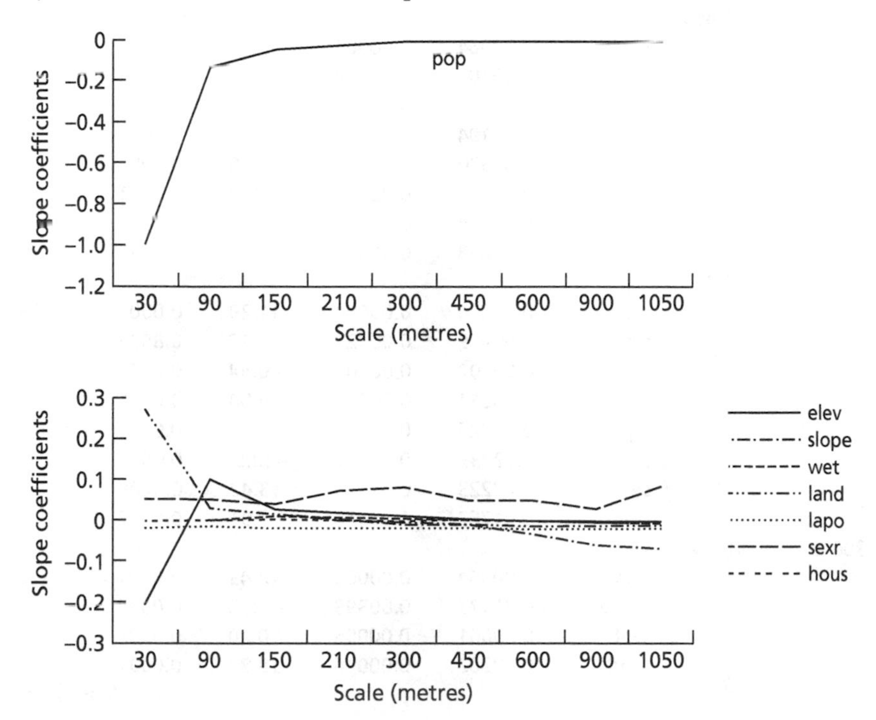

Figure 2.2 *Plots of slope coefficients for the (a) population and (b) environment variables.*

Table 2.1 *Regression summary across scales: NDVI dependent variable*

Scale (m)	*n*	*Var*	*Slope*	*Std. error*	*t-value*	$P > \|t\|$	*Adj.* R^2
30	10,532						
		elev	0.00081	0.00005	14.68	0.0001	0.100
		slope	−0.00214	0.00177	−1.21	0.2270	
		wet	0.00016	0.00004	3.72	0.0002	
		land	0.25716	0.01794	14.34	0.0001	
		lapo	−0.01931	0.00242	−7.97	0.0001	
		pop	−0.99231	0.12279	−8.08	0.0001	
		sexr	0.05193	0.02672	1.94	0.0520	
		hous	−0.20867	0.51556	−0.40	0.6857	
90	10,918						
		elev	0.00072	0.00005	13.56	0.0001	0.126
		slope	−0.00462	0.00253	−1.83	0.0679	
		wet	0.00001	0.00004	0.30	0.7673	
		land	0.02885	0.00188	15.34	0.0001	
		lapo	−0.01384	0.00218	−6.34	0.0001	
		pop	−0.13149	0.01305	−10.07	0.0001	
		sexr	0.04954	0.02506	1.98	0.0481	
		hous	0.09857	0.05598	1.76	0.0783	
150	11,456						
		elev	0.00064	0.00004	13.28	0.0001	0.148
		slope	0.00107	0.00279	0.39	0.7000	
		wet	0.00007	0.00004	1.53	0.1264	
		land	0.01194	0.00063	18.83	0.0001	
		lapo	−0.01670	0.00196	−8.50	0.0001	
		pop	−0.05071	0.00433	−11.69	0.0001	
		sexr	0.03714	0.02163	1.72	0.0859	
		hous	0.02668	0.01829	1.46	0.1449	
210	11,757						
		elev	0.00054	0.00004	11.29	0.0001	0.153
		slope	−0.00059	0.00327	−0.18	0.8576	
		wet	−0.00002	0.00005	−0.44	0.6620	
		land	0.00611	0.00031	19.54	0.0001	
		lapo	−0.01726	0.00185	−9.31	0.0001	
		pop	−0.02733	0.00217	−12.59	0.0001	
		sexr	0.07223	0.02114	3.42	0.0006	
		hous	0.01769	0.00928	1.91	0.0567	
300	8,362						
		elev	0.00064	0.00005	12.43	0.0001	0.192
		slope	−0.01173	0.00399	−2.93	0.0034	
		wet	−0.00001	0.00005	−0.20	0.8398	
		land	0.00302	0.00016	18.83	0.0001	

(*continued*)

Table 2.1 *(Continued)*

Scale (m)	*n*	*Var*	*Slope*	*Std. error*	*t-value*	$P > \|t\|$	*Adj.* R^2
		lapo	−0.01653	0.00189	−8.72	0.0001	
		pop	−0.01419	0.00112	−12.60	0.0001	
		sexr	0.08037	0.02185	3.68	0.0002	
		hous	0.01221	0.00486	2.51	0.0121	
450	3,823						
		elev	0.00061	0.00007	8.39	0.0001	0.225
		slope	−0.01579	0.00684	−2.31	0.0209	
		wet	−0.00002	0.00007	−0.28	0.7785	
		land	0.00144	0.00009	14.48	0.0001	
		lapo	−0.01594	0.00257	−6.21	0.0001	
		pop	−0.00600	0.00069	−8.59	0.0001	
		sexr	0.04644	0.02952	1.57	0.1158	
		hous	0.00209	0.00299	0.70	0.4842	
600	2,204						
		elev	0.00071	0.00009	7.86	0.0001	0.260
		slope	−0.03566	0.00939	−3.80	0.0002	
		wet	−0.00006	0.00009	−0.71	0.4766	
		land	0.00084	0.00006	12.30	0.0001	
		lapo	−0.01691	0.00307	−5.50	0.0001	
		pop	−0.00364	0.00047	−7.70	0.0001	
		sexr	0.05279	0.03404	1.55	0.1210	
		hous	−0.00186	0.00202	0.92	0.3580	
900	1,020						
		elev	0.00076	0.00013	6.00	0.0001	0.279
		slope	−0.06205	0.01507	−4.12	0.0001	
		wet	−0.00011	0.00013	−0.87	0.3869	
		land	0.00035	0.00004	8.60	0.0001	
		lapo	−0.01322	0.00391	−3.39	0.0007	
		pop	−0.00172	0.00029	−5.93	0.0001	
		sexr	0.02751	0.04785	0.57	0.5655	
		hous	0.00173	0.00130	1.33	0.1836	
1050	758						
		elev	0.00082	0.00014	6.00	0.0001	0.337
		slope	−0.06867	0.01677	−4.10	0.0001	
		wet	−0.00006	0.00014	−0.42	0.6740	
		land	0.00026	0.00003	8.33	0.0001	
		lapo	−0.01366	0.00405	−3.38	0.0008	
		pop	−0.00130	0.00023	−5.75	0.0001	
		sexr	0.08771	0.04988	1.76	0.0791	
		hous	0.00154	0.00100	1.54	0.1250	

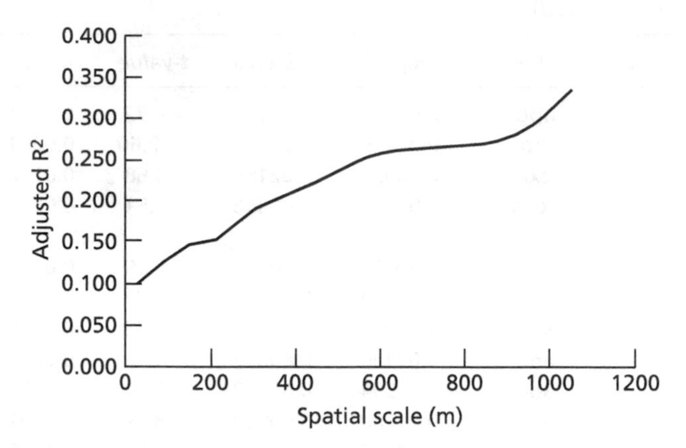

Figure 2.3 *Plot of the adjusted* R^2 *values for the NDVI models across selected spatial scales.*

Canonical Correlation

Canonical correlation was used to examine the nature of the relationship between a group of biophysical variables observed at the pixel-level (i.e., plant biomass ("NDVI"), elevation ("elev"), slope angle ("slope"), and soil wetness potential ("wet")) to a group of population variables observed at the village-level, but transformed to the pixel-level, (i.e., total land under cultivation ("land"), ratio of total land under cultivation and total population ("lapo"), total population ("pop"), ratio of males to females ("sexr"), and number of households ("hous"). The goal of the analysis was to examine the spatial and statistical variations in the relationships between groups of population and environmental variables as a consequence of the scale of observation (Walsh et al., 1999).

The hypothesis was that the relationships between environmental variables, representing biophysical site conditions and environmental gradients, and the population variables, representing the ability of the village population to alter the landscape through deforestation, reforestation, and agricultural extensification/intensification, varied collectively as a function of spatial scale. To examine possible scale dependence, canonical coefficients were derived to relate biophysical variables to environmental axes (Table 2.2), and social variables to population axes (Table 2.3). Tables 2.2 and 2.3 present the canonical coefficients for three of the selected nine spatial scales. While data were generated at each of the scale steps, only coefficients for the 30, 150, and 1,050 m scales are presented here.

Table 2.2 *Canonical coefficients for the environment variables*

At 30 m

Var	*1*	*2*	*3*
NDVI	0.2208	*0.8236*	0.1317
elev	*0.9451*	−0.5915	−0.0594
slope	0.0349	0.3784	0.7682
wet	0.0898	−0.3727	*1.0618*

At 150 m

Var	*1*	*2*
NDVI	0.3870	*−0.8530*
elev	*0.8266*	0.7246
slope	0.1018	−0.5000
wet	0.1245	0.2385

At 1050 m

Var	*1*	*2*	*3*
NDVI	*0.7337*	−0.5441	−0.3840
elev	0.0923	*1.2758*	−0.9401
slope	0.4838	−0.2601	*1.5440*
wet	0.0450	0.4845	0.2519

Note: Standardized canonical coefficients eliminate unit problem across variables

Canonical correlations indicate the number of axes in canonical space that are significant at the 0.001 level. Table 2.2 indicates that "elev" is associated with axis 1, "NDVI" with axis 2, and "wet" with axis 3 at the 30 m scale. At the 150 m scale, only two axes are statistically significant: axis 1 is related to "elev" and axis 2 is related to "NDVI." At the 1,050 m scale, three axes are statistically significant: axis 1 is associated with "NDVI," axis 2 with "elev," and axis 3 with "slope."

Table 2.3 indicates that three axes are statistically significant at the 30 m scale for the population variables: axis 1 is related to "hous," axis 2 to "land," and axis 3 to "pop." At the 150 m scale, only two axes are significant: axis 1 is related to "pop" and axis 2 is related to "hous," whereas at the 1,050 m scale, three axes are defined, axis 1 is related to "pop," axis 2 to "hous," and axis 3 to "lapo." At the 30 m scale, "lapo" is related to "elev" at

Table 2.3 *Canonical coefficients for the population variables*

At 30 m

Var	*1*	*2*	*3*
land	−0.0924	*1.9742*	0.2690
lapo	0.5798	−1.4504	−1.3019
pop	−2.5506	−0.3573	*−1.8045*
sexr	0.4429	−0.1708	0.4731
hous	*2.7585*	−1.7839	0.6913

At 150 m

Var	*1*	*2*
land	0.2147	−1.8202
lapo	0.3726	1.2873
pop	*−2.7338*	−0.1389
sexr	0.3664	0.2957
hous	2.6568	*2.1912*

At 1050 m

Var	*1*	*2*	*3*
land	0.7839	−1.2210	−1.4972
lapo	−0.0498	0.9273	*1.9552*
pop	*−2.4210*	−1.7370	1.3213
sexr	0.2590	0.5312	−0.8629
hous	1.6362	*3.4231*	−0.3748

Note: Standardized canonical coefficients eliminate unit problem across variables

axis 1; "hous" to "NDVI" at axis 2; and "land" to "elev" at axis 3. At the 150 m scale, "lapo" is related to "elev" at axis 1 and "hous" to "NDVI" at axis 2. At the 1,050 m scale, "lapo" is associated with "NDVI" at axis 1, "hous" to "elev" at axis 2, and "sexr" to "slope" at axis 3.

Results and Discussion

Multiple regression models

From a review of Table 2.1 and Figure 2.2, it is apparent that elevation ("elev"), amount of land under cultivation ("land"), the ratio of land under cultivation and population ("lapo"), and total population ("pop") each are significant at the 0.05 level in understanding the variance in plant biomass across spatial scales. In general, the variables related to topography increase in their effect as the scale becomes coarser, whereas key variables derived from the social survey decrease in their effects with scale. For example, the effects of elevation are strongest at the finest and coarsest scales; the strength and significance of slope angle increases with scale; land becomes weaker at coarser scales, though it is significant throughout the range of spatial scales sampled; population becomes weaker at coarser scales; and the ratio of land to people is fairly steady in its strength and effect over scale. Furthermore, elevation is positively correlated with plant biomass, meaning that plant biomass is higher in upland areas. This relationship is as expected, given the relative remoteness of upland areas and that the majority of nonriparian forest in the area exists on upland sites as well. Total land under cultivation is also positively correlated with plant biomass.

The significant negative effect of population, especially at finer scales, may reflect more recent settlement of upland areas where cassava is being grown. The negative relationship between plant biomass and "lapo" similarly expresses that the smaller the amount of cultivated land per person in an area, the higher the natural vegetation and thus the higher the plant biomass in February. Elevation effectively bifurcates the study area by LULC. Large expanses of rice and isolated individual and clumped patches of trees typify the lowland sites, whereas relatively large expanses of forests, interrupted by cassava and sugar cane, typify the upland sites. The topographic transition between the lowlands and uplands are characterized by low, medium, and high terraces. Along these terraces, farmers may extensify their agriculture in response to crop prices and above-average precipitation levels. February is a normally dry month and 1993 was a below-average water year. Therefore, "slope" and biomass were found to be negatively correlated. The negative relationship between slope and plant biomass at medium to coarse resolutions suggests that plant biomass does not vary significantly with "slope" at the more local scale, but regionally is associated with the flatter areas as found in lowland riparian corridors. The curvilinear pattern of the elevation coefficients may reflect riparian vegetation and vegetable gardens set close to streams offset by shifting patterns of cultivation moving up the slopes of the terraces as well as the dominant contrast of upland and lowland sites that are visible at coarser scales.

As expected, higher biomass is associated with lower total population as well as lower values of "lapo" (the ratio of land under cultivation and total population). The history of agricultural extensification in the area explains the negative relationship between population and plant biomass, since less settled or developed areas are not as exposed to agriculture and thus should have greater amounts of natural vegetation (such as forest).

A consistent finding with other scaling work is the increasing model explanatory power with increasing scale. Spatial aggregation and the smoothing of pattern through reduced image (as well as the descriptor variables) variance impacts the R^2-values across the sampled scales. At 30 m resolution, the adjusted R^2 is 0.100, whereas by 1,050 m resolution the adjusted R^2 climbs to 0.337. There is in fact a strong positive correlation (0.980) between the scale of the analysis and the amount of variance in plant biomass explained by the model. Such results point to two important considerations: (i) while elevation, area under cultivation, cultivated area per person, and total population are significant predictors across spatial scales, they explain more variance at the regional level than at the local level, and (ii) there is still a large amount of variance in plant biomass unaccounted for by the population and environment variables used here, especially at finer spatial scales. There are several possible reasons. First, plant biomass was calculated for a date in February of 1993, a month that is typically dry in a water year that experienced less rainfall than normal across the region. As such, plant biomass may have been lower that year in the areas with lower water accessibility. Future work may incorporate actual (versus potential) moisture gradients from climatological data in order to account for annual precipitation variation. Second, it may be that the variables used here would perform better for pre-rice harvest date imagery than they did for the post-rice harvest February image. However, assessing intra-annual vegetation variation is difficult with optical remotely sensed data because of the cloud cover associated with the growing season in tropical Southeast Asia.

Canonical correlations

The canonical coefficients and loadings indicate the nature of the changing relationship between a set of population variables and a set of environmental variables with scale. The defined axes were formed through a linear combination of variables from a set or pool of variables. The association of "lapo" and the "sexr" with "NDVI" appear strongest as indicated through the loadings. While results from only three of the 9-scale steps are reported here, the data supports the contention that the relationship between the population and environmental variables, as a group, are scale dependent

As reported by Walsh et al., (1999), the statistical results indicate an "upland effect" associated with the site suitability for cassava. Cassava (and sugar cane to a lesser degree) was the primary crop linked to deforested land within the district. It is generally restricted to the higher elevations and the drier sites. Depending upon the time of year and market conditions, upland crops are in various stages of development or the fields are being prepared for replanting. Plant productivity levels track relative to growth cycles, and social variables, particularly the sex ratio, are associated with villages with a significant amount of their prescribed village territory growing upland crops. Even at coarser spatial scales, the relatively large field sizes that characterize the upland crops maintain a distinctive forest-agricultural signal in the NDVI data, because of the spectral contrast of the field crops against the relatively large expanse of forest occurring as the background, landscape matrix.

In addition, the finer-scale social variables appear to be composited onto broader-scale environmental gradients. The effect of human activity, as revealed through the survey responses, is easier to see at a finer scale. Relevant decision-making units are in many cases households, which on the social side, represents a fine scale. The nature of environmental transitions of site conditions and LULC are influenced by broader-scale factors such as climate, geomorphology, and lithology that affect site conditions and land suitabilities for agriculture through local and regional resource endowments and terrain settings. Human decision-making is made within this resource-location context that influences LULC through changes in plant greenness. The relevance of scale dependent linkages between people and the landscape is what is suggested here.

Conclusions

Much remains to be accomplished in the study of scale dependence as applied to an environment strongly impacted by both human and biophysical processes. Determining important relationships across scale and defining the effects to be scaled as inputs to local, regional, and global models of LULCC are significant challenges. Initial steps include the integration of social and biophysical variables to represent drivers of LULCC, deriving statistical relationships across space and time scales, and interpreting results relative to existing theory and techniques documented in associated areas of scientific inquiry.

GIS, remote sensing, social surveys, and spatial and statistical analyses offer a research synergism suitable to addressing questions related to population and environment interactions in numerous environments, including this site in northeast Thailand. While the present research remains a preliminary study of scale dependence, the results do suggest that the relationships between biophysical and social variables change in nature and strength as the scale of examination changes. This finding is important in light of the large volume of LULC research currently underway by a variety of research teams around the world. Based on preliminary research reported elsewhere, it is clear that different teams are operating at different spatial scales. This paper suggests that the likely result of differently scaled research efforts will be different, and perhaps incompatible conclusions. A comprehensive study of population and the environment that expects to contribute to cumulative geographic thought must therefore be sensitive to scale dependent relationships, and should avoid analyses framed within a traditional single-scale perspective.

References

American Society for Photogrammetry and Remote Sensing 1998: *Proceedings, PECORA 13 Symposium, Human Interactions with the Environment: Perspectives from Space*. Bethesda, MD {ISBN-1-57083-055-X [CD]}.

Arbhabhirama, A., D. Phantumvanit, J. Elkington, and P. Ingkasuwan 1988: *Thailand Natural Resources Profile*. Singapore: Oxford University Press.

Barrows, H. 1923: Scientific human ecology as geography, *Annals of the Association of American Geographers*, 13: 1–14.

Benson, B. J. and M. D. MacKenzie 1995: Effects of sensor spatial resolution on landscape structure parameters, *landscape Ecology*, 10(2): 113–120.

Beven, K. J. and M. J. Kirkby, 1979: A physically based variable contributing area model of basin hydrology, *Hydrological Sciences Bulletin*, 24(1): 43–69.

Bian, L. and S. J. Walsh 1993: Scale dependencies of vegetation and topography in a mountainous environment in Montana, *The Professional Geographer*, 45(1): 1–11.

Blaikie, P. and H. Brookfield 1987: *land Degradation and Society*. London: Methuen.

Cousins, S. H. 1993: Hierarchy in ecology: Its relevance to landscape ecology and geographic information systems, In R. Haines-Young, D. R. Green, and S. H. Cousins (eds), *Landscape Ecology and GIS*. New York: Taylor & Francis, 75–86.

Crawford, T. W. 1999: A comparison of region building methods used to examine human–environmental interactions in Nang Rong District, Northeast, Thailand, *Proceedings, Applied Geography Conference* (Schoolmaster, F. A., ed.), 22: 366–73.

Crawford, T. W. 2000: *Human–environment Interactions and Regional Change in Northeast Thailand: Relationships Between Socio-Economic, Environmental, and Geographic Patterns*. Doctoral Dissertation, Department of Geography, University of North Carolina – Chapel Hill.

Crews-Meyer, K. A. 1999: Modeling land-cover change associated with road corridors in Northeast Thailand: Integrating normalized difference vegetation indices and accessibility surfaces, *Proceedings, Applied Geography Conference*, (Schoolmaster, F. A., ed.), 22: 407–16.

Crews-Meyer, K. A. 2000: *Integrated Landscape Characterization Via Landscape Ecology and GIScience: A Policy Ecology of Northeast Thailand.* Doctoral Dissertation, Department of Geography, University of North Carolina - Chapel Hill.

Entwisle, B., Walsh, S. J., Rindfuss, R. R., and Chamratrithirong, A., 1998: Land-use/land-cover (LULC) and population dynamics, Nang Rong, Thailand, National Academy of Science/National Research Council. *Proceedings, People and Pixels*, 121–44.

Evans, T. P. 1998: *Integration of Community-Level Social and Environmental Data: Spatial Modeling of Community Boundaries in Northeast Thailand.* Doctoral Dissertation, Department of Geography, University of North Carolina - Chapel Hill.

Forman, R. T. T. and M. Godron 1986: *Landscape Ecology.* New York: John Wiley.

Forman, R. T. T. 1995: Some general principles of landscape and regional ecology, *Landscape Ecology*, 10(3): 133–42.

Fukui, H. 1993: *Food and Population in a Northeast Thai Village.* Monographs of the Center for Southeast Asian Studies, Kyoto University, English-Language Series, No. 19, Honolulu: University of Hawaii Press.

Ghassemi, F., A. J. Jakeman, and H. A. Nix 1995: *Salinisation of Land and Water Resources: Human Causes, Extent, Management and Case Studies.* Sydney: University of New South Wales Press Ltd.

Goudie, A. et al. 1994: *The Encyclopedic Dictionary of Physical Geography*, 2nd Edition. Oxford: Blackwell Publishers Ltd.

Johnston, R. J., D. Gregory, and D. M. Smith (eds) 1994: *The Dictionary of Human Geography*, 3rd Edition. Oxford: Blackwell Publishers Ltd.

Kaida, Y. and V. Surarerks 1984: Climate and agricultural land-use in Thailand. In M. M. Yoshino (ed.), *Climate and Agricultural Land Use in Monsoon Asia.* Tokyo: University of Tokyo Press, 231–54.

Klijn, F., and H. A. Udo de Haes 1994: A hierarchical approach to ecosystems and its implications for ecological land classification, *Landscape Ecology*, 9(2): 89–104.

Lambin, E. F. 1997: Modelling and monitoring land-cover change processes in tropical regions, *Progress in Physical Geography*, 21(3): 375–93.

Liverman, D., Moran, E.F., Rindfuss, R.R., and Stern, P.C. 1998: *People and Pixels – Linking Remote Sensing and Social Science.* Washington, DC: National Academy Press.

Luque, S. S., R. G. Lathrop, and J. A. Bognar 1994: Temporal and spatial changes in an area of the New Jersey Pine Barrens landscape, *Landscape Ecology*, 9(4): 287–300.

Meentemeyer, V. and E. O. Box 1987: Scale Effects in Landscape Studies. In M. G. Turner (ed.), *Landscape Heterogeneity and Disturbance.* New York: Springer-Verlag, 15–36.

Millington, A., S. J. Walsh, and P. Osborne 2000: *Remote Sensing and GIS Applications in Biogeography and Ecology.* Norwell, MA: Kluwer Academic Publishing.

Moody, A. and Woodcock, C. E. 1995: The influence of scale and the spatial characteristics of landscapes on land-cover mapping using remote sensing, *Landscape Ecology*, 10(6): 363–79.

Moore, I. D., Grayson, R. B., Landson, A. R. 1991: Digital terrain modeling: A review of hydrological, geomorphological, and biological applications, *Hydrological Processes*, 5: 3–30.

Nassauer, J. I. 1995: Culture and changing landscape structure, *Landscape Ecology*, 10(4): 229–37.

Parnwell, M. J. G. 1988: Rural poverty, development and the environment: The case of Northeast Thailand, *Journal of Biogeography*, 15: 199–313.

Parnwell, M. J. G. 1992: Confronting uneven development in Thailand: The potential role of rural industries, *Malaysian Journal of Tropical Geography*, 22(1): 51–62.

Quattrochi, D. A., and M. F. Goodchild (eds) 1997: *Scale in Remote Sensing and GIS*. New York: Lewis Publishers.

Qi, Y. and J. Wu 1996: Effects of changing spatial resolution on the results of landscape pattern analysis using spatial autocorrelation indices, *Landscape Ecology*, 11(1): 39–49.

Risser, P. G., Karr, J. R., Forman, R. T. T. 1984: Landscape ecology: Directions and approaches. *Illinois Natural History Survey, Special Publication 2*, Champaign.

Rundel, P. W. and K. Boonpragop 1995: Dry forest ecosystems of Thailand, In Bullock, S. H., H. A. Mooney, and E. Medina (eds), *Seasonally Dry Tropical Forests*. New York: Cambridge University Press, 93–123.

Turner, M. G., 1990a: Landscape changes in nine rural counties in Georgia, *Photogrammetric Engineering and Remote Sensing*, 56(5): 379–86.

Turner, M. G. 1990b: Spatial and temporal analysis of landscape patterns, *Landscape Ecology* 4(1): 21–30.

Urban, D. L., O'Neill, R. V., and Shugart, H. H. 1987: Landscape ecology: A hierarchical approach can help scientists understand spatial patterns, *Bioscience*, 37(2): 119–27.

Vink, A. P. A. 1983: *Landscape Ecology and Land Use*. New York: Longman.

Walsh, S. J., Butler, D. R., Allen, T. R., and Malanson, G. P. 1994a: Influence of snow patterns and snow avalanches on the alpine treeline ecotone, *Journal of Vegetation Science*, 5: 657–72.

Walsh, S. J. and K. A. Crews-Meyer, 2001: *Remote Sensing and GIS Applications for Linking People, Place, and Policy*. Norwell, MA: Kluwer Academic Publishers.

Walsh, S. J., F. W. Davis, and R. K. Peet (eds) 1994b. Applications of remote sensing and geographic information systems in vegetation science, *Journal of Vegetation Science*, 5: 641–56.

Walsh, S. J., Evans, T. P., Welsh, W. F., Entwisle, B., and Rindfuss, R. R., 1999: Scale-dependent relationships between population and environment in Northeast Thailand, *Photogrammetric Engineering and Remote Sensing*, 65(1): 97–105.

Walsh, S. J., Evans, T. P., Welsh, W. F., Rindfuss, R. R., and Entwisle, B. 1998: Population and environmental characteristics associated with village boundaries and land-use/land-cover patterns in Nang Rong District, Thailand, *Proceedings, PECORA 13 Symposium, Human Interactions with the Environment: Perspectives from Space*, American Society for Photogrammetry and Remote Sensing, Bethesda, MD (ISBN-1-57083-055-X [CD Publication], 395–404.

Walsh, S. J., Moody, A., Allen, T. R., and Brown, D. G. 1997: Scale dependence of NDVI and its relationship to mountainous terrain, In D. A. Quattrochi, and M. F. Goodchild, (eds), *Scale in Remote Sensing and GIS*, Boca Raton, FL: CRC Lewis Publishers, 27–55.

3 Crossing the Divide: Linking Global and Local Scales in Human–Environment Systems[1]

William E. Easterling and Colin Polsky

Scale[2] is a human construct that locates an observer/modeler relative to a set of objects distributed in space, time, and magnitude. It explains nothing in and of itself, but its perspective may influence the discovery of pattern and process (Wilbanks and Kates, 1999; Gibson et al., 2000). Wiegleb and Broring (1996) note that simply magnifying or de-magnifying the resolution or extent of the data may make homogeneity out of heterogeneity and vice versa. Relationships that depend on the scale of analysis in this way are said to exhibit *scalar dynamics* (Turner et al., 1995; Geoghegan et al., 1998). The purpose of this chapter is to examine the role of scalar dynamics in models of complex human–environment interactions.

We begin with a synopsis of scale problems in models of human–environment interactions. Next we examine a set of theoretical issues that frame these scale problems in terms of complex systems dynamics. A full account of methods for modeling these theoretical notions is beyond the scope of this chapter. As a starting point, we review how a technique that is well understood by many scholars – statistical regression analysis – may be modified for use in empirical studies of scalar dynamics. Standard regression techniques do not explicitly distinguish among the scales that may be implicated in a process, but a relatively new technique called multilevel (or hierarchical) modeling is designed for this purpose. With this technique, information from multiple scales is incorporated into a single regression model. We present a hypothetical multilevel model of land-use, and briefly discuss two recent studies using this technique. We conclude by drawing the theoretical and methodological topics together.

The Issue of Scale in Models of Human–Environment Interactions

The current understanding of global environmental change has emerged from several decades of intense investigation of whole earth systems. The establishment of long-term global observing systems coupled with massively parallel computing have led to large leaps in understanding planetary scale processes. However, in empirical terms the linkage of those processes to finer-scale regional and local processes remains a challenge (Wilbanks and Kates, 1999). In theoretical terms, the deep analysis of the meaning of scale in global environmental change studies called for by Clark (1985) is still emerging as a coherent area of research (e.g., O'Neill, 1988; Root and Schneider, 1995; Kohn, 1998; Easterling and Kok, 2002). To date, issues of scale in the global environmental change literature mostly concerns practical modeling problems of aggregating in situ modeling experiments into large-scale projections or, conversely, down-scaling macro relationships to meso and micro scales (Harvey, 2000).

Such issues are captured, respectively, in the "bottom-up" and "top-down" paradigms that typify global environmental change modeling, especially regarding the simulation of biophysical and human response to climate change (Root and Schneider, 1995). The reference point for both paradigms is spatial and temporal scale. Bottom-up modeling is typified by process-level simulation of biophysical response to a change in climate variables across a range of "representative" modeling sites. In some cases, the results are recursively linked to an integrative (usually economic) model of the region or globe to produce results with policy relevance (e.g., Rosenberg, 1993; Rosenzweig and Parry, 1994; Parry et al., 1996). A top-down approach is typified by reduced-form relations between climate, biophysical, and socioeconomic variables that are estimated from aggregate (usually country-level) data to estimate large-scale (often global) impacts of environmental changes (Alcamo et al., 1994; Nordhaus, 1992; Dowlatabadi, 1995; Edmonds et al., 1995). Relationships are estimated between physical variables (e.g., temperature, rainfall), biological variables (e.g., crop yields, forest growth), and socioeconomic variables (e.g., income, institutional structures).

To estimate the impacts of climate change, models from both the bottom-up and top-down paradigms typically incorporate results from related but independent studies. In these cases, mismatches in resolution between systems being modeled are likely. This case is illustrated best when results from large-scale general circulation model (GCM) experiments used to simulate climate change with a resolution of hundreds of kilometers are combined with results from site-specific process models with resolutions of a few meters. Such mismatches call into question the reliability of the simulated impacts, as the relationships between climate and other variables differ between the scales (Easterling et al., 2001; Mearns et al., 2001). There is a growing realization that the importance of such scalar dynamics affects more than the reliability (or lack thereof) of model results. Evidence for scalar dynamics also suggests a need to improve the sophistication of our understanding of the unit of analysis, i.e., coupled human–environment systems (Kates and Clark, 1999).

Coupled Human–Environment Systems

We use the term "coupled human–environment system" (Kates and Clark, 1999) because although it may be intuitively appealing to think of social and natural systems as distinct, the two systems are in fact inseparable. Material and energy exchange so freely between ecosystems and economic systems as to make boundaries between them indistinguishable except by convention (for example, the boundary between the market and nonmarket shown in Figure 3.1). Social systems consume energy-laden, low-entropy natural materials (e.g., biomass, fossil fuels) for self-maintenance and the production (or reproduction) of new material forms (Ayres, 1994). This consumption transforms the low-entropy material into high-entropy heat and material (waste) that is subsequently deposited back into the natural system (Ayres, 1998). It is therefore not possible to conceive of a natural system that operates completely independently of social systems.

To understand how natural and human systems are coupled requires an additional dimension to that of energy and material flows: information flows. For example, the flow of genetic information is crucial to the evolution and survival of both human and natural systems (Kohn, 1998). Human systems also persist because they actively store social information in the form of formal and informal social institutions (Kohn, 1998). These institutions facilitate social learning across generations, which has a self-organizing effect (van der Leeuw, 2001). Thus as human systems organize – for example, movement from a hunter-gatherer to an economically specialized society – they increase in complexity over time. In the long run these processes necessarily increase entropy for the system as a whole (Ayres, 1998). Thus it is fair to say that the organization of society serves to increase disorder in natural systems. Hence our definition of the term coupled human–environment systems: spatiotemporal assemblages of ecosystems, their abiotic controls (climate mostly), and the socioeconomic systems that

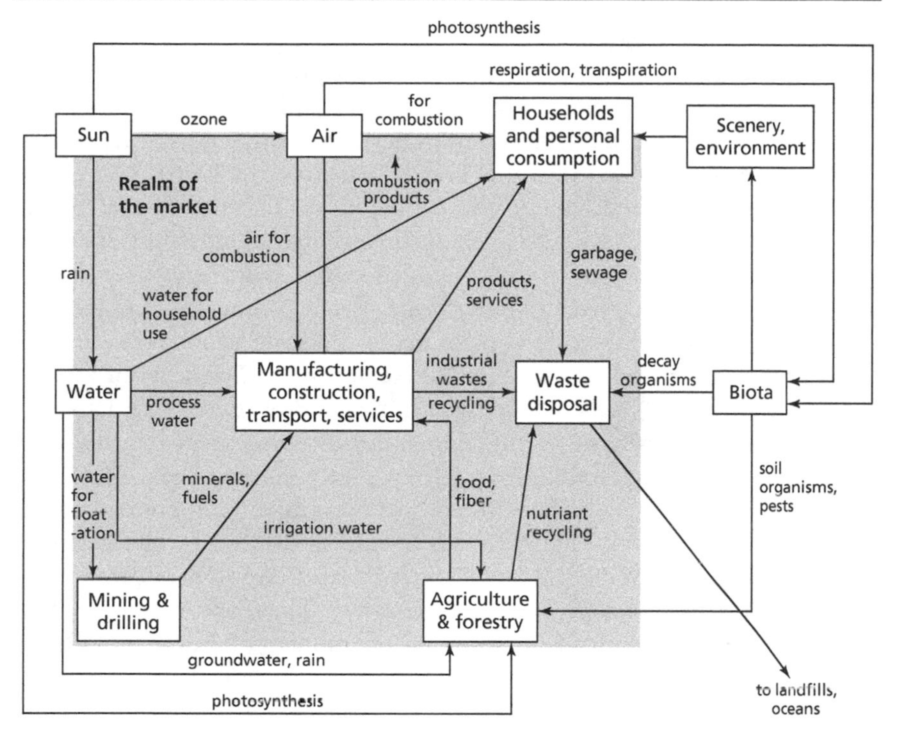

Figure 3.1 *The complex human–environment system.*
Source: After Ayres (1994).

derive benefit from ecosystems and interact with them to promote entropy system-wide.

Several scholars have argued convincingly that coupled human–environment systems are structured hierarchically in space and time (Allen and Starr, 1982; O'Neill, 1988; Müller, 1996; Kohn, 1998; Perrings, 1998; Buenstorf, 2000). A hierarchy is a partially ordered set of objects ranked according to asymmetric relations among themselves (Shugart and Urban, 1988). Descriptors useful in distinguishing levels of a hierarchy include, for example, larger/smaller than, faster/slower than, to embed/to be embedded in, and to control/to be subject to control. For example, in ecosystems the behavior of lower levels in the hierarchy (e.g., individual organisms) is explained by biological mechanisms such as photosynthesis, respiration, and assimilation (Levin, 1992). At higher levels, these biological mechanisms are relatively unimportant for understanding system behavior. Instead, abiotic processes such as climate variability and biogeochemical cycling are the most important factors to model, and may be considered constraints on the lower-level biological mechanisms. In economic systems, the lower levels of the hierarchy (e.g., individual firms) are well understood in terms of rapidly changing production functions. The higher levels in the system – slower-moving national and international features such as rates of inflation, policy development, and national income – impose constraints on individual firms (Gibson et al., 2000).

Hierarchy theory embodies the foregoing ideas. This theory evolved out of general systems thinking to explain the multitiered structure of certain types of production systems. In simplified terms, hierarchy theory posits that the most useful way to understand the behavior of a multiscaled complex system at one scale is to examine the scale of interest plus two others, one larger and one smaller (O'Neill, 1988). The level of interest (Figure 3.2, Level 0) is a component of a higher level (Level +1). Level +1 dynamics are generally slower moving and greater in extent than Level 0; they form boundary conditions that serve to constrain the behavior of Level 0. Level 0 may also be divided into constituent components at the next lower level (Level −1). Processes operating at Level −1 are generally faster moving and smaller in spatial extent than Level 0; they provide the mechanisms that explain Level 0 behavior. They are represented as state variables (dynamic driving forces) in models of Level 0 (O'Neill, 1988).

Thus, the goal of hierarchy theory is to understand the behavior of complex systems by structuring models to capture dynamics at a small number of interrelated scales. Many scales, of course, may be implicated in a given process, but hierarchy theory posits that system dynamics may be satisfactorily captured through a judicious selection of three levels. Hierarchy theory provides an important framework for global environmental change studies by establishing a strong rationale for modeling processes at multiple scales simultaneously. This approach is designed to simplify some of the considerable complexity of human–environment systems.

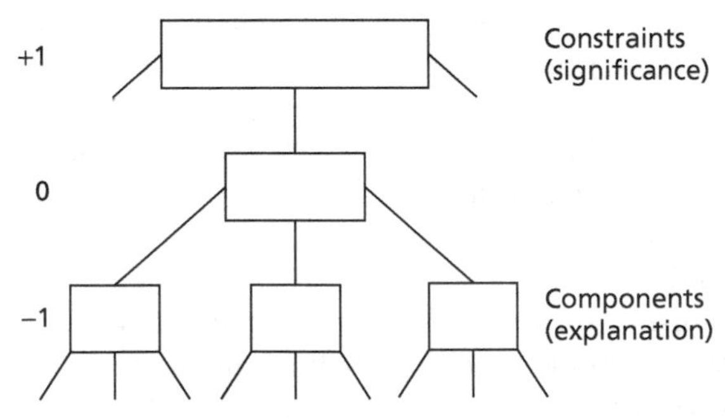

Figure 3.2 *Levels of a hierarchy.*
Source: After O'Neill (1988).

Complexity and Coupled Human–Environment Systems

Why does it matter that processes driving a system are organized hierarchically? In other words, why do some properties of a system appear at certain scales and disappear at others? A theoretical explanation for the existence of hierarchically ordered relations in coupled human–environment systems requires unraveling the very meaning of complexity. We can only touch on basic concepts here but refer the reader to Prigogine and Stengers (1984); Manson (2001); and Reitsma (2003) for a more complete accounting of the principles of complexity theory.

As noted above, complex coupled human–environment systems are dissipative structures in a far-from-equilibrium state.[3] Buenstorf (2000) argues that such systems must import low-entropy (energy and material) and export high-entropy (heat) into the environment. Energy and material fluxes are nonlinear in a far-from-equilibrium state. This nonlinearity is illustrated in coupled human–environment systems by the frequently observed drive toward decreasing national energy intensities (expressed as unit energy input per unit Gross Domestic Product). Even as GDP increases in absolute terms in many postindustrial economies on a nonlinear trajectory, energy intensity begins a nonlinear decline. Several researchers argue these nonlinearities to be a necessary condition for self-organization.

Self-organization is the dynamic emergence of structures and properties at the whole-system level that are developed through intense interactions between system components. Buenstorf (2000) concludes that self-organization is driven by random fluctuations in energy and material fluxes that result in bifurcations – large and unpredictable jumps – in system outputs. Prigogine and Stengers (1984) posit this process to be "order through fluctuations."

We illustrate these complicated ideas of self-organization, hierarchy, and bifurcations with the case of technological change in US agriculture over the past half-century. Although normally evaluated by its large-scale effects on production, the drive toward technological innovation is inescapably a regionalized process. Hayami and Ruttan (1985) posit that technological innovation in agriculture is induced endogenously. According to their "induced innovation hypothesis," as inputs become scarce, prices rise. Persistent high prices convey strong signals to the agricultural research establishment to develop new technologies that substitute for more expensive and older inputs, to reduce production costs. Because regional variations in resource endowments lead to regional differences in farmers' comparative advantage, the pattern of induced innovation will be likewise regionally distributed. Farmers at one location have quite different sets of technological needs than those at another location.

The development of successful hybrid corn varieties illustrates the point. In the US, each state land grant (agricultural) university has its own corn breeder. This regionalized approach to supporting research on corn hybridization represents the "invention of a method of inventing" corn varieties, adapted to each growing region (Hayami and Ruttan, 1985). The successful development and diffusion of commercial hybrid corn varieties has been accomplished by the evolution of a complex research, development, distribution, and educational system. This system has depended on close regional cooperation among public sector research and extension agencies, a variety of public, semipublic, and cooperative seed-producing organizations, and private sector research and marketing agencies.

This regionalized innovation system persists by producing a steady stream of fine-tuning adjustments that adapt cropping systems to challenges in their local production environments (i.e., spatial and temporal variability in pests, climate, soils) – as long as producer and input (factor) prices do not exceed critical thresholds. Occasionally, however, in response to unusual trends in factor prices, one technical innovation (or set of innovations) rises above others in importance. The innovation is so important that it rapidly diffuses in space and time, dramatically altering system productivity. The rapid increase in productivity may amount to a phase shift or, as noted above, a bifurcation. Following the logic of Prigogine and Stengers (1984), if the bifurcation in output is large enough, new macro institutional structures may emerge that constrain lower level behavior. For the coupled human–environment system of agriculture, this is usually accomplished through government programs and prices (Witt, 1997). The application of nitrogen fertilizers to corn and eventually to virtually all row crops begun shortly after World War II in the US, coupled with the hybridization of corn (described above), brought about a remarkable upsurge in yields and production. From a system-wide perspective, such major innovations probably had the initial appearance of a random series of regional fluctuations in farming practices (technological innovations). Eventually, these regional fluctuations became so pervasive that the entire system had fundamentally changed through nonlinear relationships, i.e., positive or negative feedbacks.

Buenstorf (2000) argues that positive feedback from high levels of a system is necessary to amplify random fluctuation at low levels in order to create structure of higher levels. In the US, the Cooperative Extension Service, through its hierarchical bureaucracy, serves an amplifier (positive feedback mechanism) by directing appropriate technologies to large numbers of farmers. Negative feedback is also required to stabilize interactions between low-level components. Crop prices can act as high-level negative feedback mechanisms in market-driven production systems. A surge in production (the aforementioned bifurcation) prompted by technological innovations (low-level random fluctuation) without a corresponding increase in demand leads to lower prices. Global agricultural output prices, adjusted for inflation, have been on a downward trajectory for several decades. This trajectory has discouraged reckless farming practices that damage the resource base while inducing further technological innovations aimed at minimizing production costs.

The Challenge of Modeling Coupled Human–Environment Systems

From the practical standpoint of policy-motivated modeling, the failure to match temporal and spatial scales of human activities with those of nature has been an abiding problem in the science of climate and society interactions (Clark, 1985; Holling, 1994). This failure stems in part from the misinterpretation of modelers at the whole system level of the meaning of self-organizing pulses of change upwelling from finer scales in the system and the emergent structures that these pulses create.

Modeling pulses of change across multiple scales of a coupled human–environment system is a difficult challenge. Most simple systems consisting of a small number of elements can be understood structurally and modeled mechanistically (Figure 3.3, Region I). Such systems represent organized simplicity. The full description of a simple two-object system requires only four equations: one for each object to describe how the object behaves by itself ("isolated" behavior equation), one to describe how the behavior of

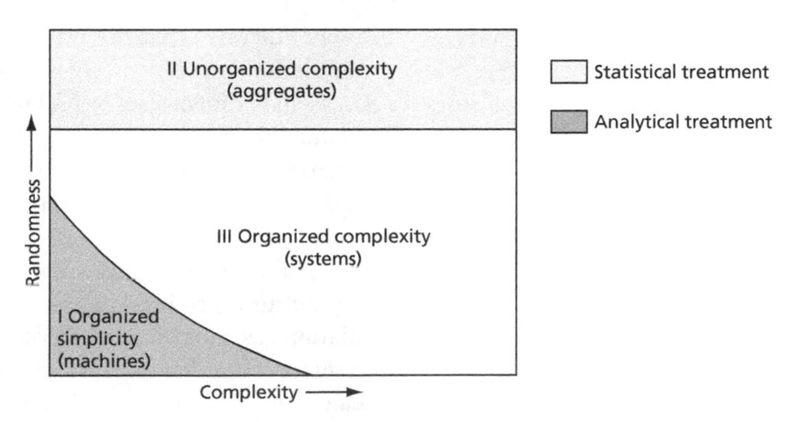

Figure 3.3 *Complexity versus randomness.*

each object affects that of the other ("interaction" equation) and one to consider how the system behaves absent the objects ("field" equation). As the number of objects increases, the number of interaction equations increases by the "square law of computation" (2^n, where n is the number of objects). For example, modeling a 10-object system requires 1,035 equations: one for the field equation, ten for the individual objects, and $2^{10} = 1,024$ for the interactions. Complex human–environment systems consist of many times more than 10 objects.

There is a paradox in the above paragraph. Small-scale simplicity does not necessarily equate to mechanistic behavior of a system. At sufficiently small scales, under nonlinear external forcing even deterministic systems can exhibit chaotic behavior. It was this revelation that gave rise to quantum physics, which seeks to understand small-scale systems in terms of unorganized complexity (Figure 3.3, Region II).

It is an open question as to whether coupled human–environment systems are purely deterministic at any level of scale, even at a scale in the domain of organized simplicity. This poses a challenge to modeling human–environment interactions even at the scale of individual agents. But it is possible to simulate generalized human behavior at small scales as a simple stochastic system with a finite number of possible outcomes, using agent-based modeling and other stochastic approaches. This is analogous to simulating organized simplicity in Figure 3.3. The complexity of interactions rises with scale, however, because the number of agents increases with scale. An agent-based model that tracks every agent's interactions with every other agent, and with the environment, rapidly eludes comprehension and computation even with massively parallel processing.

Yet, at the extreme of large numbers of agents, Weinberg (1975) notes that the interactions within certain populations may be random and therefore predictable in a statistical sense by aggregating to high levels of scale. In such populations, the "law of large numbers" dictates that the probability that a property of any one object in the population will deviate significantly from the average value of that property across all N objects is $1/\sqrt{N}$. Hence, the larger the value of N, the more predictable the property becomes. According to Weinberg (1975) such populations are complex but random in their behavior such that they can be studied statistically – they represent unorganized complexity (Figure 3.3, Region II). Of course taking this approach sacrifices the ability to model feedbacks in highly interactive systems such as farms within a national agricultural production system.

Some complex systems with large populations cannot be characterized as exhibiting organized simplicity or unorganized complexity. Systems with relationships that are random at one scale but organized at other scales exhibit organized complexity (Figure 3.3, Region III). Land-use change modeling, a mainstay of efforts to understand changes in the Earth's biogeochemical cycles, illustrates this point. At large spatial scales (e.g., countries, groups of countries), Turner and Meyer (1991) argue that Ehrlich and Holdren's (1971) IPAT identity usefully explains large-scale patterns and trends of land-use change. IPAT is defined as defined as: Intensity of human impact on the environment [I] = Population [P] × Affluence [A] × Technology [T]. Elements of IPAT are easily identified in global integrated assessment models such as the IMAGE 2.0 model (Alcamo et al., 1994). Such models typically simulate land-use change as a function of change in agricultural demand, which is approximated by changes in population and per capita income. The difficulty with the IPAT identity is that the dynamics of coupled human–environment systems can be characterized as neither organized simplicity nor unorganized complexity.

The behavior of coupled human–environment systems can be described as in the domain of organized complexity. At small spatial scales (e.g., households, villages), Turner and Meyer (1991) posit that many driving forces influence land-use and land-cover change, including economics, culture, and politics, in addition to the environment. For example, in some cultures a change in the structure of familial inheritance of land may be as important in the determination of land-use as changes in land rent. Such features are difficult to model and are therefore typically absent from large-scale models of land-use change. In general, environmental impacts appear to correlate well with only a few factors – income, population, technology – at large spatial scales (e.g., countries, groups of countries) but poorly at small spatial scales (e.g., farm households, villages) where a wide range of factors such as cultural mores and local governance structures are more relevant. This variation in the importance of explanatory factors between scales highlights the possibility of potentially serious prediction errors if the analyst fails to account for the variation (Turner and Meyer, 1991).

Studying Scalar Dynamics Using Statistical Regression

The theoretical issues outlined above suggest that an analysis of organized complexity (e.g., coupled human–environment systems) should account for the possibility that different relationships dominate at different levels of organization. In methodological terms, there does not appear to be a best approach for evaluating these subtleties. A reasonable methodological base from which to approximate how complex human–environment systems operate is statistical regression analysis. Regression is perhaps the most commonly used multivariate quantitative technique in all of social science and as such is well understood by a wide audience of scholars. However, regression as commonly employed is not well suited for examining scalar dynamics. A relatively new technique, *multilevel* (or *hierarchical*) modeling is designed for gaining insight into the operation of hierarchically ordered systems. As such this technique provides more credible information on scalar dynamics. For this reason we turn now to a discussion of how regression analysis – suitably modified – may be employed profitably in the study of multiscaled, complex human–environment systems.

Standard regression models (typically referred to as Ordinary Least Squares, or OLS models) are not designed to examine scalar dynamics. Indeed, an implicit assumption in OLS models is that all variables "operate" at the same scale. Thus even comparing multiple regression models where data are aggregated to different scales between the models cannot adequately capture scalar dynamics. Calculating results for each scale in isolation from the other scales is not consistent with the mandate from hierarchy theory discussed above. Hierarchical organization suggests that system behavior depends on the simultaneous influence of factors from multiple scales. It stands to reason that scalar dynamics cannot be effectively assessed one scale at a time.

Multilevel (or *hierarchical*) *modeling* is a regression technique that evaluates scale-specific relationships at multiple scales considered jointly. Between-scale interactions are modeled explicitly. Instead of specifying all independent variables at all scales, the scale(s) at which the variables operate is explicit in how the variables are specified. This single equation, multiple scale approach therefore evaluates inter-scalar relationships directly. The important feature of multilevel models is that estimates of estimated regression coefficients and associated variances are *scale-specific*.

Consider a hypothetical study of US agricultural land-use where (for simplicity) only two factors are considered: a farmer's tillage practice as a small-scale influence, and state-scale land conservation policy as a large-scale influence. In this case the nested hierarchy of spatial scales hypothesized to influence the process of land-use is farmers, nested within states. All farmers are subject to a policy effect, but the policy effect should vary between states.[4] An OLS study of this process must, by construction, specify the tillage and policy effects at the same level (either average values for tillage would be aggregated to the state-level, or the state-level policy values would be assigned to all farm-level observations). Such an approach would encounter the scale-mismatch problem discussed above. Aggregating farm-level values for tillage necessarily obscures the within-state variation on this measure. As a result, valuable information is lost and the precision of the associated relationships is diminished. Disaggregating the state-level values for policy implies that the policy is associated with farm-level behavior, when in fact the policy process is largely independent of farm-level decision-making. As a result, information is effectively (if accidentally) "created" and the precision of the associated relationships is overstated.

Alternatively, we argue that it is preferable to partition the scale-specific effects and interactions explicitly according to the appropriate scales. In this way precision is stated with greater accuracy (which is not to say maximum accuracy – see below). A hierarchical model is designed to account for these scale-specific differences. We identify six aspects of hierarchical model results that have relevance for a scalar dynamics study. A hierarchical regression model of this hierarchy will estimate separate effects for:

1 large-scale policy;
2 small-scale tillage;
3 a cross-scale tillage-policy interaction effect;
4 the small-scale variance of land-use;
5 the large-scale variance of land-use;
6 the large-scale variance of the small-scale tillage effect.

All of these effects are estimated with explicit reference to the specified spatial hierarchy, which is defined *a priori*. None of these effects can be estimated in this way using OLS in a multiscale context (although there are similarities – see below). We elaborate on these assertions in the next section.

The added value of this single equation, multiple scale hierarchical approach compared to the multiple equation, multiple scale OLS approach discussed in the preceding section is evident from inspecting the general form of the hierarchical model one scale at a time.[5] We use the hypothetical example above of farmers nested within states as motivation, but the structure is general. The model is first introduced one scale at a time for purposes of exposition. Then we combine the equations representing the different scales into the integral single-equation model.

The first, or lowest, level in the hierarchy is the farm-scale. Assume for simplicity that at the farm-level, only one variable, tillage (TILL), predicts land-use (L-USE),

$$\text{L-USE}_{ij} = \beta_{0j} + \beta_{1j}\text{TILL}_{ij} + r_{ij} \quad (1)$$

where i indexes the farms (level −1 units) and j indexes the states (level −2 units). Now let the intercepts and slopes from equation (1) – the β_{0j} and β_{1j}, respectively – be expressed as dependent variables in another set of regressions for the state-scale, each with a randomly varying error term. In this way analysts estimate the extent to which farm-level behavior varies between states (without explaining such variation). As such the hierarchical model estimates a separate intercept and slope for each of the states:

$$\beta_{0j} = \gamma_{00} + \mathbf{u}_{0j}, \tag{2}$$

$$\beta_{1j} = \gamma_{10} + \mathbf{u}_{1j}, \tag{3}$$

Combining equations (2) and (3) into (1) results in a simple hierarchical model with predictor variables at the smallest (farm) scale only. This specification is equivalent to the basic random coefficients model common in econometrics.

Assuming there is significant between-state variation in farmer behavior, the next step is to explain the variation. This step requires the extension of the state-scale equations (2) and (3) to include state-level explanatory variables, in this hypothetical case, state land conservation policy (POLICY):

$$\beta_{0j} = \gamma_{00} + \gamma_{01}\mathrm{POLICY}_j + u_{0j}, \tag{4}$$

$$\beta_{1j} = \gamma_{10} + \gamma_{11}\mathrm{POLICY}_j + u_{1j}, \tag{5}$$

Combining equations (4) and (5) into (1) results in the simplest version of a full two-level hierarchical model with predictor variables at each level:

$$\begin{aligned}\mathrm{LUSE}_{ij} &= \gamma_{00} + \gamma_{01}\mathrm{POLICY}_j + \gamma_{10}\mathrm{TILL}_{ij} + \gamma_{11}\mathrm{POLICY}_j\mathrm{TILL}_{ij} \\ &\quad + (u_{0j} + u_{1j}TILL_{ij} + r_{ij}),\end{aligned} \tag{6}$$

The various error terms (collected within the parentheses in equation 6) are assumed to be normally distributed with zero mean, nonzero covariance within scales, and zero covariance between scales (Bryk and Raudenbush, 1992).

The six dimensions of added insight alluded to above can be seen directly from equation (6):

1 the state-scale policy effect on farm-scale land-use is given by γ_{01};
2 the farm-scale tillage effect on farm-scale land-use is given by γ_{10};
3 the cross-scale tillage-policy interaction effect (i.e., how state-scale policies affect the farm-scale relationship between tillage and land-use) is given by γ_{11};
4 the within-state variance of farm-scale land-use is given by the variance of r_{ij};

5 the between-state variance of land-use controlling for state-scale policy is given by the variance of u_{0j};
6 the between-state variance of the farm-scale tillage effect controlling for state-scale policy is given by the variance of u_{1j}.

The first three elements in this list are associated with the main effects (regression coefficients) of the model, and could be approximated using the standard OLS, i.e., nonhierarchical, approach. OLS models provide unbiased estimates of regression coefficients even when the data are structured hierarchically (Bryk and Raudenbush, 1992). However, the precision of the estimates for these effects will be misleading. When a system is organized into a hierarchy the values for the variables within levels are systematically associated. Such nonrandom relationships diminish the effective sample size compared to the reported sample size in OLS, because OLS models assume independence among observations (Snijders and Bosker, 1999). This conflict between the necessarily nonrandom relationships within levels in hierarchical systems and the assumption of independence in OLS in turn implies that the "true" standard error of the estimate is larger than the reported standard error. As a result, the *t*-statistics will be inflated, so the probability of rejecting a true null hypothesis will be larger than reported. Thus ignoring the nested structure of the data, relationships will be estimated with greater confidence than is warranted.

The last three elements in this list cannot be estimated using the OLS approach. OLS models are constructed with a single global error term. Hierarchical models partition the overall unexplained variation into its scale-specific components. For example, the simple hierarchical model above has a three-part error term. The result is a more detailed analysis of contextual effects compared to the OLS approach, which collapses all unexplained variation into a single error term. This distinction underscores an additional advantage to the hierarchical approach: the scale-specific variances from multiple hierarchical models can be compared to calculate whether the addition of a predictor variable significantly reduces the variation *for that scale* rather than for the overall model (Bryk and Raudenbush, 1992). In short, hierarchical models allow for a targeted scale-specific decomposition of the unexplained part of the model.

Hierarchical modeling is emerging as a technique for spatial analyses of coupled human–environment interactions. For example, Polsky and Easterling (2001) estimate a hierarchical model to assess influences on agricultural land values in the US Great Plains at two scales simultaneously, the *county* and the *district* (collections of counties). Similarly, Hoshino (2001) estimates a hierarchical model to study influences on the distribution of Japanese farmland at two scales simultaneously, the *municipality* and the *prefecture* (collections of municipalities). For both of these hierarchical models, results indicate significant variation in the small-scale (county, municipality) land-use relationships across the large-scale units (districts, prefectures), which are then modeled as a function of large-scale characteristics. Polsky and Easterling (2001) find that the small-scale relationship between mean July maximum temperatures and land values varies directly with large-scale interannual temperature variability. Hoshino (2001) finds, among other things, that accounting for large-scale labor market conditions improves the explanatory power of local-scale farm-size on the percentage of land devoted to farmland. Thus both studies show how specific large-scale factors can condition small-scale relationships. This evidence is consistent with the notion of a hierarchical spatial effect, or scalar dynamic, that analysts should expect given the theoretical concerns deriving from organized complexity discussed above. In principle, hierarchical models may be extended to include any number of levels (Goldstein, 1995).

Conclusion

Is the effect of scale in the modeling and analysis of complex human–environment systems to introduce new "substance" to the system or is it simply an illusion to the eye of the modeler? Some analysts posit scale to be more than simply the context in which state variables operate. Instead, scale is viewed as forming constraints on system behavior. That is, scale helps to explain how objects in a system interact collectively and individually. Others view scale as nothing more than an artifact of the positioning of the observer/modeler with respect to the system of interest. At one resolution the observer/modeler sees the trees only, and at another resolution the observer/modeler sees the forest only (Easterling and Kok, 2003). We see merit to both views and suggest a synthesis.

Complex systems theory provides a useful context for evaluating and understanding the dynamic role of scale in modeling coupled human–environment systems. Complex coupled human–environment systems are dissipative structures in a far-from-equilibrium state. Energy and material fluxes in such systems tend to be nonlinear, which are necessary ingredients of self-organization. Self-organization is the dynamic emergence of structures and properties at the whole-system level that are developed by strong nonlinear interactions between system components. One mechanism that drives self-organization in human–environment systems is local or regional fluctuations in energy and material fluxes that result in bifurcations – large and unpredictable jumps – in system outputs. The role of technical change in prompting large increases (bifurcations) in agricultural production in the last half-century in the United State illustrates this point. We argue that large-scale institutions such as markets and government support programs were influenced by and in turn provided important self-reinforcing feedback to the process of regional technological changes. This argument implies a hierarchical structure.

We conclude that studying hierarchically structured systems demands models that are dynamic enough to examine variable-process relations across a range of temporal and spatial scales. This conclusion presents a tall order for any modeling exercise. In this chapter we present a relatively simple regression-based approach called multilevel (or hierarchical) modeling as a way to estimate relationships at multiple scales simultaneously. Using this technique the hierarchical structure posited by hierarchy theory can be modeled explicitly. The simple example given above – where agricultural land-use is related to factors at two spatial scales and one temporal scale (point in time) – can be easily extended to more spatial scales and multiple points in time.

We find multilevel models considerably more attractive than the standard OLS regression approach, where all variables are implicitly assumed to "operate" at the same scale. However, caution is warranted because no regression approach can capture all the dynamics and dimensions of complex, coupled human–environment processes. At small scales in particular, these systems exhibit organized complexity, i.e., they are too unpredictable to be analyzed completely in statistical terms. That said, we feel the statistical paradigm will remain an important domain for empirical research on coupled human–environment systems.

In particular, although the error term in statistical models contains a lot of information, there are few studies that explicitly examine spatial and temporal patterns of the error term, at least within the global environmental change literature.[6] To the extent that statistical models (whether multilevel or not) fail to capture system dynamics, examining "poorness-of-fit" may highlight important factors missing from the original model. Finally, it should be noted that all quantitative models will suffer from limited explanatory power because some of the important factors in the processes under study cannot be quantified (see discussion above on IPAT). Thus modeling exercises such as those critiqued and discussed in this chapter should be used in an iterative and complementary fashion with qualitative studies (e.g., interviews of farmers and agricultural scientists that may shed light on the non-IPAT factors important to land-use).

Modeling coupled human–environment systems requires deep thinking about how scale influences what we know and how we know it. Novel approaches are required for dealing with highly nonlinear variable-process relations across time and space in such systems. Multilevel modeling represents an important step towards realizing this goal. We expect other approaches to emerge and be blended in the coming decades.

Notes

1 Portions of this paper draw extensively from Easterling and Kok (2002 in press).
2 We use Gibson et al.'s (2000) taxonomy of scale-related terms for this paper. *Scale* refers to the spatial or temporal dimensions used to measure phenomena. *Extent* is the size of the scale. *Resolution* is the precision of measurement of objects of a scale. A *level* consists of objects that share the same extent and resolution.
3 Equilibrium state in this case refers to thermodynamic equilibrium in which maximum entropy is the steady state, i.e., "heat death" prevails.
4 Such hierarchical spatial effects are a special case of systematic spatial association, or spatial autocorrelation. In the present example, the (state-scale) policy variable is (necessarily) spatially autocorrelated for farmers in the same states. This chapter does not address issues of spatial autocorrelation in detail. Statistical models that explicitly account for spatial autocorrelation are not new in geography (e.g., Cliff and Ord, 1973) or in studies of agricultural land-use (e.g., Haining, 1978). Recent developments in this field are summarized by Anselin (2001), and applied to the case of land-use by, among others, Hsieh (2000), Munroe (2001), and Polsky (2002).
5 In this section we rely on the discussion and notation of multilevel models presented in Bryk and Raudenbush (1992). Full technical details of hierarchical models may be found in that text as well as in Jones, 1991; Goldstein, 1995; Hox, 1995; Kreft and DeLeeuw, 1998; Snijders and Bosker, 1999.
6 For an example of this approach, see the analysis of recent US agricultural land values by Polsky (2002).

References

Alcamo, G. J., J. Kreileman, J. S. Krol, and G. Zuidema 1994: Modeling the global society-biosphere-climate system: Part I: Model description and testing. *Water, Air and Soil Pollution*, 76: 1–35.

Allen, T. F. H. and T. B. Starr 1982: *Hierarchy-Perspectives for Ecological Complexity*. Chicago, IL: University of Chicago Press.

Anselin, L. 2001: Spatial Econometrics. In B. Baltagi (ed.), *Companion to Theoretical Econometrics*. Oxford: Blackwell Publishers.

Ayres, R. U. 1994: Industrial metabolism: theory and policy. In B. R. Allenby and D. J. Richards (eds), *The Greening of Industrial Ecosystems*. Washington, D.C.: National Academy Press, 23–37.

Ayres, R. U. 1998: Eco-thermodynamics: Economics and the second law. *Ecological Economics*, 26: 189–209.

Bryk, A. S. and S. W. Raudenbush 1992: *Hierarchical Linear Models: Applications and Data Analysis Methods*. Sage, Beverly Hills, CA: University of California Press.

Buenstorf, G. 2000: Self-organization and sustainability: energetics of evolution and implications for ecological economics. *Ecological Economics*, 33: 119–134.

Clark, W. C., 1985: Scales of Climate Impacts. Climatic Change, 7: 5–27.

Cliff, A. D. and J. K. Ord, 1973: Spatial Autocorrelation. Pion, London.

Dowlatabadi, H. 1995: Integrated assessment models of climate change. *Energy Policy*, 23(4/5): 289–96.

Easterling, W. E. and K. Kok, 2003: Emergent properties of scale in global environmental modeling – are there any? In J. Rotmans and M.V. Asselt (eds), *Scaling in Integrated Assessment,* Lisse: Sets and Zeitlinger, 263–92.

Easterling, W. E., C. Polsky, D. Goodin, M. W. Mayfield, W. A. Muraco and B. Yarnal 2001: Analyzing greenhouse gas emission inventories at multiple spatial scales in the US. In R. Kates and T. Wilbanks (eds), *Global Change and Local Places: Estimating, Understanding, and Reducing Greenhouse Gases*. Cambridge: Cambridge University Press.

Edmonds, J. A., D. Barns, M. Wise and M. Ton 1995: Carbon coalitions: The cost and effectiveness of energy agreements to alter trajectories of atmospheric carbon dioxide emissions. *Energy Policy*, 23(4–5): 309–36.

Ehrlich, P. R. and J. P. Holdren 1971: Impact of Population Growth. *Science*, 171(3977): 1212–17.

Geoghegan, J., L. Pritchard, Y. Ogneva-Himmelberger, R. K. Chowdhury, S. Sanderson and B. L. Turner 1998: "Socializing the Pixel" and "Pixelizing the Social" in Land-Use and Land-Cover Change. In D. Liverman, E. F. Moran, R. R. Rindfuss and P. C. Stern (eds), *People and Pixels: Linking Remote Sensing and Social Science*. Washington, DC: National Academy Press.

Gibson, C., E. Ostrom and T.-K. Ahn 2000: The concept of scale and the human dimensions of global change: a survey. *Ecological Economics*, 32: 217–39.

Goldstein, H. 1995: *Multilevel Statistical Models*. Kendall's Library of Statistics 3. London: Edward Arnold.

Haining, R. P. 1978: A spatial model for high plains agriculture. *Annals of the Association of American Geographers*, 68(4): 493–504.

Harvey, L. D. D. 2000: Upscaling in global change research. *Climatic Change*, 44: 225–63.

Hayami, Y. and V. Ruttan 1985: *Agricultural Development: An Agricultural Perspective*. Baltimore, MD: The Johns Hopkins University Press.

Holling, C. S. 1994: Simplifying the complex: the paradigms of ecological function and structure. *Futures*, 24(6): 598–609.

Hoshino, S. 2001: Multilevel modeling on farmland distribution in Japan. *Land Use Policy*, 18: 75–90.

Hox, J. 1995: *Applied Multilevel Analysis*. Amsterdam: TT-Publikaties.

Hsieh, W. 2000: Spatial dependence among county-level land-use changes. Unpublished PhD Dissertation Thesis, The Ohio State University, Columbus, Ohio.

Jones, K. 1991: Specifying and estimating multilevel models for geographical research. *Transactions of the Institute of British Geographers*, 16: 148–59.

Kates, R.W. and W. C. Clark (eds) 1999: *Our Common Journey: A Transition Toward Sustainability*. Washington, DC: National Academy Press.

Kohn, J. 1998: Thinking in terms of system hierarchies and velocities: What makes development sustainable? *Ecological Economics*, 26: 173–87.

Kreft, I. G. G. and J. DeLeeuw, 1998: *Introducing Multilevel Modeling*. London: Sage Publications.

Levin, S. 1992: The problem of pattern and scale in ecology. *Ecology*, 73(6): 1943–67.

Manson, S. 2001: Simplifying complexity: A review of complexity theory. *Geoforum*, 32(3): 405–14.

Mearns, L. O., W. E. Easterling, C. Hays and D. Marx, 2001: Comparison of agricultural impacts of climate change calculated from high and low resolution climate model scenarios: Part I. The Uncertainty of Spatial Scale. *Climatic Change*, 51: 131–72.

Müller, F. 1996: Emergent properties of ecosystems: consequences of self-organizing processes? *Senckenbergiana maritima*, 27(3/6): 151–68.

Munroe, D. 2001: Patterns of small-scale farm production in Poland: scale dependence in a spatial econometric framework. Unpublished PhD Dissertation Thesis, The University of Illinois, Urbana-Champaign, Illinois.

Nordhaus, W. 1992: An optimal transition path for controlling greenhouse gases. *Science*, 258: 1315–19.

O'Neill, R. V. 1988: Hierarchy theory and global change. In T. Rosswall, R. G. Woodmansee and P. G. Risser (eds), *SCOPE 35, Scales and Global Change: Spatial and Temporal Variability in Biospheric and Geospheric Processes*. Chichester: John Wiley and Sons, 29–45.

Parry, M. L., J. E. Hossell, P. J. Jones, T. Rehman, R. B. Tranter, J. S. Marsh, C. Rosenzweig, G. Fischer, I. G. Carson and R. G. H. Bunce 1996: Integrating global and regional analyses of the effects of climate change: A case study of land-use in England and Wales. *Climatic Change*, 32: 185–98.

Perrings, C. 1998: Resilience in the dynamics of economy-environment systems. *Environmental and Resource Economics*, 11(3–4): 503–20.

Polsky, C. 2002: A spatio-temporal analysis of agricultural vulnerability to climate change: The US Great Plains, 1969–1992. Unpublished PhD Dissertation Thesis, The Pennsylvania State University, University Park, Pennsylvania.

Polsky, C. and W. E. Easterling 2001: Adaptation to climate variability and change in the US Great Plains: A multiscale analysis of Ricardian climate sensitivities. *Agriculture, Ecosystems, and Environment*, 85(1–3): 133–44.

Prigogine, I. and I. Stengers 1984: *Order out of chaos. Ecology*. New Science Library, Boulder, CO.

Reitsma, F., 2003: A Response to Simplifying Complexity. *Geoforum*. 34(1): 13–16.

Root, T. L. and S. H. Schneider 1995: Ecology and climate: Research strategies and implications. *Science*, 269: 334–41.

Rosenberg, N. J. 1993: Towards an integrated impact assessment of climate change: The MINK study. *Climatic Change*, 24: 1–173.

Rosenzweig, C. and M. L. Parry 1994: Potential impact of climate change on world food supply. *Nature*, 367: 133–8.

Shugart, H. H. and D. L. Urban 1988: Scale, synthesis, and ecosystem dynamics. In L. R. Pomeroy and J. J. Alberts (eds), *Concepts of Ecosystem Ecology: A Comparative View*. New York: Springer-Verlag, 23–37.

Snijders, T. A. B. and R. J. Bosker 1999: *Multilevel Analysis: An Introduction to Basic and Advanced Multilevel Modeling*. New York: Sage.

Turner, B. L. and W. B. Meyer 1991: Land use and land-cover in global environmental change: Considerations for study. *International Social Science Journal*, 130: 669–79.

Turner, B. L., D. L. Skole, S. Sanderson, G. Fischer, L. Fresco and R. Leemans (eds), 1995: land-use and land-cover change: Science/research plan. Joint publication of the International Geosphere-Biosphere Programme (Report No. 35) and the Human Dimensions of Global Environmental Change Programme (Report No. 7). Royal Swedish Academy of Sciences, Stockholm.

van der Leeuw, S. E. 2001: "Vulnerability" and the integrated study of socio-natural phenomena. *International Human Dimensions Program Update*, 01(2): 6–7.

Weinberg, G. M. 1975: *An Introduction to General Systems Thinking*. New York: Wiley.

Wiegleb, G. and U. Broring 1996: The position of epistemological emgergentism in ecology. *Senckenbergiana maritima*, 27(3/6): 179–93.

Wilbanks, T. J. and R. Kates 1999: *Global Changes in Local Places: How Scale Matters*. *Climatic Change*, 43: 601–28.

Witt, U. 1997: Self-organization and economics-what is new? *Structural Change Economic Dynamics*, 8: 489–507.

4 Independence, Contingency, and Scale Linkage in Physical Geography

Jonathan D. Phillips

There is a certain subdisciplinary schizophrenia in physical geography on the issue of scale linkage (transferring information, relationships, models, and rules between different spatial and temporal scales). On the one hand there has long been an intuitive recognition that processes and environmental controls relevant at a given spatial or temporal scale may be of little or no relevance at much broader or narrower scales. On the other hand, many physical geographers have implicitly if not explicitly accepted a reductionist notion that one may, at least in theory, start with first principles at the most detailed level and from there move seamlessly up the scale hierarchy to the broadest level. As empirical evidence and formal arguments began to accumulate to bolster the view that the "rules" change as scales change, some geoscientists began to seriously question whether it is even theoretically possible or worthwhile to attempt seamless representations across the entire range of scales relevant to geography and earth science. More or less simultaneously, disappointment and frustration with the lack of success in scale linkage achieved by the reductionist approach led to the same general questions (Baker and Twidale, 1991; Beven, 1995; Church, 1996; Imeson and Lavee, 1998; Larson and Kraus, 1995; Sherman, 1995). Even as the tools and methods for transferring models and information between disparate scales multiply and improve, the suspicion grows that there may be fundamental limits to the range of spatiotemporal scales across which seamless representations are possible.

The *American Heritage Dictionary* contains numerous definitions of "scale." The two closest to the geographic conception are "a progressive classification, as of size, amount, importance, or rank; and "a relative level or degree." In physical geography the concept of scale is relatively straightforward and generally concerns either geographic extent/longevity or spatial/temporal resolution, or a hierarchical organization or level of generalization of environmental systems. Physical geography deals with scales from the molecular to the planetary; from the instantaneous to the geological. It is therefore axiomatic that scale linkage is a critical concern for physical geography and indeed for geosciences in general. For example, the relationship between erosional unloading and isostatic adjustment is important in the study of fluvial geomorphology, as is the relationship between flow hydraulics and sediment transport mechanics. However, the former relationship operates over broad spatial scales and geological time scales; the latter at the level of individual grains and essentially instantaneously. Both relationships are governed by well-understood geophysical principles. If the ultimate goal of fluvial geomorphology is to derive a comprehensive understanding of fluvial systems, we are confronted with the question of how relationships operating at such disparate scales can be accomodated in the same theoretical or pedagogical construct.

This chapter is concerned with two key aspects of scale linkage. The *operational* problems of scale linkage relate to the extent to which phenomena which operate or vary over different spatial and temporal scales can be or should be treated independently in the analysis of earth surface systems. For instance, in the study of stream systems, can relationships between erosion and isostasy, and between flow hydraulics and sediment transport, be considered independent? May one safely examine landscape evolution involving fluvial downcutting and compensatory isostatic uplift without taking account of sediment transport mechanics? The *contextual* problems of scale linkage concern the extent to which a landscape or earth surface system must or should be interpreted in the context of environmental controls operating at different spatial and temporal scales. For instance, can we develop a generally applicable model based on sediment entrainment and transport laws to describe channel incision downstream of dams? Or can such a model be applied only with due attention to the specific situation of each dam site and the time frame involved?

Operational issues will be discussed first, with a focus on whether processes or environmental factors operating at different scales are independent. A case will be made that a seamless representation from the most detailed scales to the broadest is not possible in physical geography, but that such scale linkage can be accomplished across a limited range of the relevant scale domains. This finding is then discussed in the context of contextual scale linkage. It will be argued that operational difficulties with scale linkage do not reduce or eliminate the necessity of interpreting results and relationships in the context of broader and more detailed scales.

Operational Issues of Scale Linkage

Few will argue that the controls over process–response relationships often vary with spatial and temporal scale. The question is whether processes or environmental factors operating at fundamentally different scales are independent. If, say, a single transport event for a sand grain and dune system evolution are independent, then either can be confidently studied without including, or even accounting for, the other. The more fundamental question is whether laws at any scale from grain to dune field can be extended to both ends of the continuum.

Intuitive arguments

Intuition provides some powerful suggestions that scale linkage across the entire range of geographically relevant scales is at least unfeasible, if not impossible. The classic paper of Schumm and Lichty (1965) addresses this indirectly by suggesting that geomorphic variables which are dependent at one time scale may be independent at others. There is no literature, for example, even attempting to account for landscape-scale soil geography (much less global soil distributions) on the basis of the molecular level processes which are clearly important in other aspects of soil science. Is there any conceivable way one might explain Quaternary climate change on the basis of the detailed behavior of gas molecules? Or a way to bring plate tectonics to bear on the problem of the relative importance of specific soil erosion processes?

Common sense suggests that different methods, techniques, and modes of inquiry will be required as temporal and spatial scales change. At the operational level, scale-appropriate methods must be used, and linkage across the full range of scales is unlikely. However, intuition also suggests that it will be necessary in many cases to embed or interpret results in the context of some other (usually broader) scale. Thus, while plate tectonics does not contribute to the explanation of erosion mechanics, in some cases the detailed erosion rates and processes in a given location are best interpreted in the context of their tectonic setting.

Several empirical studies have shown directly (and many more indirectly or implicitly) that the controls over process–response relationships change as temporal and spatial scales change. A sampling is given in Table 4.1. In Table 4.2, a variety of more general arguments for this principle in the geosciences are summarized.

Table 4.1 *Selected empirical studies showing how controls over process–response relationships differ with scale (an illustrative, not exhaustive list)*

Reference	*Phenomenon*
Blandford, 1981	Soil erosion and sediment yield on rangelands
Braun & Slaymaker, 1981	Snowmelt runoff
Chappell et al., 1996	Hillslope soil redistribution processes
Imeson & Lavee, 1998	Soil erosion–climate relationship
Larson & Kraus, 1995	Cross-shore sediment transport on sandy beaches
Penning-Roswell & Townsend, 1978	Stream channel slopes
Phillips, 1986b	Marsh shoreline erosion
Poesen et al., 1994	Effect of rock fragments on soil erosion by water
Reed et al., 1993	Environmental controls on vegetation
Seyfried & Wilcox, 1995	Runoff response to precipitation
Smit, 1999	Impact of grazing on soil humus

Hierarchy theory

Hierarchy theory has been proposed as a conceptual framework for scale linkage in geomorphology and ecology (DeBoer, 1992; Haigh, 1987; O'Neill, 1988; O'Neill et al., 1986). This is sometimes seen as a mechanism which can facilitate a seamless up- or downscaling linkage. However, the implications of hierarchy theory in fact presuppose the impossibility or difficulty of a seamless linkage, and focus on the search for appropriate scales or hierarchical levels and the principles operative at those levels. Given a space-time scale (level i), hierarchy theory holds that a higher-level system (i+1) provides the constraints and a lower-level system (i–1) provides the mechanistic explanation. Levels higher than i+1 are too large and slow to be observed at i and can be ignored. Levels less than i−1 are too fast and small and appear as background noise.

Hierarchy theory, then, holds that scale linkage must occur using scale-appropriate methods and principles which vary up and down the hierarchy.

Information criterion

No representation linking two processes or phenomena operating at different rates or scales should allow changes to be propagated through space faster than they actually occur in time. This is the premise of the information criterion, variants of which are discussed by Martin (1993) and

Table 4.2 *Theoretical or generalized arguments or synthetic discussions indicating scale independence*

Reference	*Phenomenon or subject*	*Summary of argument*
Beven, 1995	Hydrological modeling	Applying small-scale models to larger scales is an inadequate approach due to the dependency of hydrologic systems on historical and geological perturbations
Campbell, 1992	Fluvial erosion and sediment yield	Extrapolation from small area/short time studies is difficult because of spatiotemporal complexity and heterogeneity at broader scales precludes synthesis of smaller-scale fluxes
Cambers, 1976	Coastal erosion	Dependent/independent status of variables changes with temporal scales
Douglas, 1988	Hillslope evolution modeling	Assumption of scale independence is critical for modeling hillslope evolution
Hoosbeek & Bryant, 1992	Pedology	Separate models and approaches required at different hierarchical scales
Schumm & Lichty, 1965; Phillips, 1986a; 1988	Fluvial geomorphology; geomorphic systems	Dependent/independent status of variables changes with temporal scales; factors operating over distinctly different scales are independent
Sherman, 1995	Processes, form, and evolution of coastal dunes	Different models necessary at different scales; upscaling from process mechanics to evolutionary scale is not yet possible
Turner et al., 1993	Ecological equilibrium in landscapes	Qualitatively different landscape dynamics are associated with different spatial and temporal scales; equilibrium is a scale-dependent concept

Phillips (1995, 1997a). Given two phenomena, one of which varies at a faster time scale than the other, the critical necessary information is the rate of the two processes. Given this information, one can determine the proper relationship between the temporal resolution or time steps of a model and the spatial resolution or cell size of a model, based on the simple rule given above. This framework can be used to determine whether disparate scales can be reconciled in a single representation.

For instance, it has been suggested that the deterministic uncertainty apparent in the local scale spatial variation in the depth of B horizons in podzolized dune soils is related to unstable wetting fronts (fingered flow) in percolating water (Pereira and Fitzpatrick, 1998; Phillips et al., 1996). The problem is in reconciling the temporal scales of moisture fluxes with those of soil development.

At the spatial scale of a pedon the spatial resolution or cell size is fixed. The information criterion then shows that the ratio of temporal resolutions should be given by the rate of vertical moisture flux divided by the rate of soil formation. The rate of vertical moisture flux in the sandy soils in the Phillips et al. (1996) study area in coastal North Carolina may exceed 50 $cmhr^{-1}$, and is at least 15 $cmhr^{-1}$. The solum is everywhere at least 2 m thick, and the soils are about 77 Ka. This suggests a mean rate of conversion of dune sand into a Spodosol of about 2.96×10^{-7} $cmhr^{-1}$. Even with a 3 m solum, and soils formed in less than half the time since the dunes were deposited (say, 35 Ka), the average rate of soil formation is 9.78×10^{-7}. The narrowest plausible ratios are on the order of 15 million. Such a large ratio indicates that the dynamics of moisture flux and the development of eluvial-illuvial soil horizon relationships, while undoubtedly related, simply cannot be linked using any single model or representation based on either soil moisture dynamics or Quaternary pedogenesis – unless a model, experiment, or method of field interpretation can be devised which can reconcile time steps differing by seven orders of magnitude or more.

It may be practicable to link complex soil morphology and wetting front instability using representations based on processes or relationships at intermediate time scales. Some possibilities include the feedback effects linking illuvial accumulations to the depths of moisture penetration (Phillips et al., 1996) and the gradual leaching of water-repellent substances in flow fingers (Ritsema et al., 1998).

Physical geography contends with contemporary processes with velocities on the order of 10^{-2} to 102 $msec^{-1}$ and with historical and geological processes with velocities $\leq 10 - 11$ $msec^{-1}$. While many geographical problems do not imply large information criterion ratios, the information criterion shows that in an operational context scale linkage across all relevant scales is not feasible due to difficulties of including rates differing by several orders of magnitude in a single representation.

Abstracted systems

An earth surface system consists of interacting components which can (in principle at least) be represented as a nonlinear equation system or interaction matrix. Schaffer (1981) proved that when a system so defined includes sets of components where the temporal scale of operation or variation differs substantially, those components are independent in the sense that the slow-acting components have no effect on the stability or evolution of the system at the fast scale, and vice-versa (the abstracted systems principle). This means that in determining the system's stability properties the components operating at a given scale can be considered separately from those operating at fundamentally different scales. In physical geography the key idea is that components of earth surface systems that vary over distinctly different scales (two orders of magnitude or more has been proposed as a rule of thumb) are independent of each other in terms of their effects on system-level behavior. This argument is described in detail and applied to problems in geomorphology elsewhere (Phillips, 1986a; 1988; 1995; 1997a; 1999a).

The fundamental implication of the abstracted systems principle in physical geography is: to the extent that problems involve sets of interacting components which vary over fundamentally different scales, said components are independent of each other in terms of their influence on the evolution and stability of the landscape or earth surface system. The fluvial geomorphologist, then, may treat isostatic adjustments without accounting for particle-level erosion mechanics, and the biogeographer may examine species ranges with no concern for cell biology.

Instability and chaos

Many earth surface systems are divergent and inherently dynamically unstable and deterministically chaotic (not all, and not always, of course). This indicates that fundamental system-level behaviors vary with scale or resolution, dictating different approaches.

Landscapes often diverge over time in the sense that they become ever more differentiated into distinct spatial elements. Relief-increasing topographic evolution is an example. Dissection, uplift, or a combination results in a landscape increasingly separated into more distinct elements. The differentiation of more-or-less uniform parent material into distinct soil horizons and weathering zones is another example. Still others include some cases of ecological succession, where communities with a single dominant species are replaced by more diverse communities. A landscape or earth surface system developing in this divergent manner is dynamically unstable and chaotic. Formal demonstrations of this argument are given elsewhere, by Rosenstein et al., (1993) and Phillips (1999a).

The divergence in chaotic systems derives from deterministic underlying dynamics (in physical geography this is likely to be accompanied by sources of noise deriving from environmental controls external to the system). Chaos means, however, that at some more detailed level there must be fundamental deterministic processes at work. Chaotic behavior is also inherently limited and occurs within definite, well-defined bounds. This is intuitively consistent with the recognition that topographic relief, soil horizonation, ecological diversity, etc. cannot increase indefinitely. Thus, in a chaotic system there must be some broader scale whereby divergence and complexity are resolved into more orderly patterns (Church, 1996; Phillips, 1999a).

A canonical example of deterministic chaos is turbulent fluid flows. The underlying dynamics are based on straightforward, well-known physics. However, with more than a few particles the mechanics of fully developed turbulence must be approached probabilistically, or via nonlinear dynamical systems methods that explicitly represent the deterministic chaos. When one scales up from the details of turbulence to the aggregate behavior of fluid masses we can predict air and water flows. Thus there are at least three scale ranges where fundamental system behavior differs: the subchaotic scale of deterministic mechanics, the unstable/chaotic scale, and the broader scale defined by the inherent limits on divergent evolution. Because divergent, chaotic earth surface systems have at least three scale-related domains where system behavior differs qualitatively, scale linkage across the entire range of scales is inhibited.

Scale independence

As the arguments above indicate, the primary operational issue in scale linkage is scale independence – that is, processes and controls and relationships at different scales may be independent of each other. Once it is determined that two or more disparate scales are not independent, or the range of scales (some subset of the total range) over which seamless linkages are feasible, methods for accomplishing scale linkage are developing and evolving rapidly. These methods are generally based on explicit recognition that representations may need to be modified as scales and resolutions change (for example, Beven, 1995; Sherman 1995). There are also some emerging theories of earth surface systems which attempt to account for scale-independence by explicitly recognizing the simultaneous presence and operation of independent controls operating at different scales (Johnson et al., 1990; Huggett, 1995; 1997; Pahl-Wostl, 1995; Phillips, 1999a). Finally, application of some of the fundamental tools of spatial analysis of landscape structure can facilitate spatial scale linkage (e.g., Walsh et al. 1998).

One implication of scale independence is that fundamental, "first" principles are not constant across scales. Equating first principles with process mechanics at the most detailed scale reflects a traditional reductionist view of science. A view cognizant of scale independence recognizes that what constitutes first principles varies with spatial and temporal scale and does not necessarily correspond to microscale process mechanics. The principles governing landscape evolution or species distributions (for instance) differ from those describing erosion mechanics or plant physiology. While reductionists (if one may caricature groups of scholars) have been reluctant to accept the notion of fundamental limits on the ability to use single representations across scales, this is not necessarily so for geoscientists in general. Proponents of historical and systems-oriented approaches have acknowledged that their methods are often insufficient and sometimes useless, and have generally recognized and accepted that historical or systems approaches are best-suited to particular ranges of time and space and ill-suited to others (Baker, 1996a; 1996b; Harrison, 1999; Huggett, 1995; Pahl-Wostl, 1995). There are, therefore, traditions in geosciences which deal both explicitly with scale independence.

Contextual Issues

Practical difficulties in scale linkage and principles of scale independence might tempt some physical geographers to focus exclusively on narrow spatiotemporal scales without consideration of broader or more detailed scales. This would be a mistake. On an analytical level it may not be impossible to link different scales in a single model or representation. On an interpretive level, however, it may be necessary to consider outcomes at any given scale (or place and time) in the context of quite different scales. The necessity of scale linkage for interpretation is often an outcome of historical and spatial contingency.

Historical and spatial contingency imply that the state of an earth surface system is in part dependent on a specific history, historical event, or chain of events. Contingency has long been a key issue in human geography, and in biogeography where the role of disturbances has been widely studied.

Contingency is also emerging as a critical issue in geomorphology (Baker, 1996a; Baker and Twidale, 1991; Begin and Schumm, 1984; Harrison, 1999; Magilligan et al., 1998; Twidale, 1998), hydrology (Beven, 1995; Fan and Bras, 1995; Seyfried and Wilcox, 1995), and physical geography in general (Phillips, 1999b).

In many landscapes and environmental phenomena there is indeed dependence on factors localized in space and/or time. These local factors must be accounted for. This may involve examination of more detailed scales, as when the particular characteristics of place must be invoked to explain variations at a broader scale. For example, studies of the downstream geomorphic effects of dams in the US Great Plains show no predictable, generalizable patterns. Rather, the specific fluvial context of each river/dam determines or at least influences the response (Friedman et al., 1998; Williams and Wolman, 1984). This includes spatial, environmental factors such as hydrologic regimes, geologic setting, vegetation, etc. It also includes temporal factors such as the time since dam closure and rates of channel responses and recovery. Local contextualization may also involve upscaling to broader levels. For instance, Nicholson (2000) showed that explaining drought in the Sahel region requires placing local climates and land-surface climate interactions in the context of regional and global synoptic patterns.

Historical and spatial contingency are not necessarily relevant or important for every problem in physical geography, but contingency is both present and critical for many problems. This means that contextual (as opposed to operational) scale linkage is often necessary. To put it another way, addressing a particular research problem will generally require identifying the appropriate scale or range of scales and using methods and resolutions suited for that scale, due to scale independence. Applying the results of the study, attempting to generalize them, or even interpreting the results, however, will typically require consideration of a broader or narrower context. Consider my own studies of geomorphic changes downstream of an east Texas dam (Phillips, 2001). The results can be reasonably applied or generalized only by incorporating them in a broad spatiotemporal context (e.g., one set of evidence relevant to US Gulf and Atlantic coastal plain streams with drainage areas of 200 to 300 km^2 20 to 30 years following dam construction) or by carefully accounting for the local details of another impounded stream relative to those of my study area.

Formal and functional scales

Scale linkage, whether for contextual or analytical purposes, presupposes a hierarchy of scales, where the latter may be defined in terms of magnitude or level of generalization. This raises the question of how such hierarchies are to be defined or identified. Some scale hierarchies may be objectively defined, such as the individual, population, community, and ecosystem levels in biogeography. Scale hierarchies may also be subjectively defined, as when one refers to micro, meso, or macroscales.

Thus the definition of scale hierarchies is analagous to the definition of formal and functional regions (Juillard, 1962). Formal regions are defined on the basis of some shared geographical characteristics and are always subjective to some degree, even though quantitative or logical rules may be employed to delineate them. Functional or nodal regions are based on flows and may be objectively defined to the extent flow patterns are known. In many cases regional delineations are explicitly related to a hierarchical structure (Juillard, 1962).

The US Department of Agriculture's ecoregions are an example of formal regions and a formal scale hierarchy. The ecoregions are defined based on a combination of topography, climate, vegetation, soils, and fauna (Bailey, 1995). The ecoregions are also defined at eight hierarchical resolutions from domains which cover millions of square kilometers to land units covering tens of hectares. Just as the boundaries of the ecoregional units are somewhat subjective and often are cartographic generalizations of transition zones, the size ranges of the hierarchical ecological regions sometimes overlap. Drainage basins are an example of both functional regions and a functional scale hierarchy. The boundaries of drainage basins can be precisely, unambiguously, and objectively defined based on observed moisture fluxes or fundamental hydrologic principles. Because of the nested hierarchical structure of stream channel networks, a scale hierarchy (first-order basins, second order, . . . , etc.) can also be defined precisely, unambiguously, and objectively.

Discussion and Conclusions

Not all problems involve scale linkage, and scale linkage is not always difficult or problematic. However, because physical geography and geosciences encompass phenomena ranging from molecular to planetary, scale linkage will often be an issue. The most fundamental question is under what circumstances phenomena operating at different spatial and temporal scales can or should be considered dependent or independent.

Operationally, in the sense of attempting to transfer results or representations between scales, there are inherent fundamental limits to the range of scales across which such transfers can occur. Intuitive arguments, bolstered with empirical demonstrations, show that in some cases the factors controlling process–response relationships vary with spatial and temporal scale. Hierarchy theory holds that such transfers can take place only between adjacent hierarchical levels. The information criterion shows that in many cases it is simply not feasible to link processes or factors which vary at distinctly different rates or ranges. The abstracted systems principle provides a formal demonstration that system components operating at distinctly different scales are independent in terms of their effects on system-level dynamics. And the presence of dynamical instability and deterministic chaos in some earth surface systems points to at least three scale-defined domains of qualitatively different system behaviors.

The arguments discussed above point to an inability to transfer representations at any given scale across the entire hierarchy of interest to physical geography. They also point to the necessity of seeking methods and principles appropriate to a given scale. This implies a somewhat isolated, scale-restricted approach. This applies only to the operational issues, however. With respect to intepreting and contextualizing results, the implications are quite different – i.e., some degree of scale linkage is necessary. Another way of stating the case is that in obtaining results, seamless representations across the whole range of relevant scales is impossible and scale independence must be accounted for. This does not necessarily apply to generalizing and interpreting results, however, which may require embedding the results within the context of broader or more detailed scales.

The necessity of contextual scale linkage for interpreting results arises from the importance of historical and spatial contingency in many earth surface systems. General laws or principles may be incapable of providing explanation unless local details (lower level of a scale hierarchy) are accounted for, and/or unless the context of broader controls (higher level of a scale hierarchy) is considered.

In quantitative spatial analysis there has been a transformation from efforts to derive global laws to efforts to describe and explain spatial variability. It might be argued that an analagous transformation is underway in our efforts to address scale. Rather than attempting to construct a single representation or set of principles from particle to planet, we find it necessary to focus on understanding the variable behavior and structure of landscapes and earth surface processes at different scales.

References

Bailey, R. G. 1995: *Description of the Ecoregions of the United States*. Washington: US Department of Agriculture, Forest Service Misc. Pub. 1391.

Baker, V. R. 1996a: Hypotheses and geomorphological reasoning. In B. L. Rhoads and C. E. Thorn (eds), *The Scientific Nature of Geomorphology*. New York: John Wiley, 57–86.

Baker, V. R. 1996b: Discovering earth's future in its past: Palaeohydrology and global environmental change. In J. Branson, A. G. Brown, and K. J. Gregory (eds), *Global Continental Changes: The Context of Palaeohydrology*, London: The Geological Society, 73–84.

Baker, V. R., and Twidale, C. R. 1991: The re-enchantment of geomorphology, *Geomorphology*, 4: 73–100.

Begin, Z. B., and Schumm, S. A. 1984: Gradational thresholds and landform singularity: significance for Quaternary studies. *Quaternary Research*, 27: 267–74.

Beven, K. 1995: Linking parameters across scales: subgrid parameterizations and scale dependent hydrological models. *Hydrological Processes*, 9: 507–25.

Blandford, D. C. 1981: Rangelands and soil erosion research: a question of scale. In R. P. C. Morgan (ed.), *Soil Conservation: Problems and Prospects*. New York: John Wiley, 105–21.

Braun, L. N., Slaymaker, H. O. 1981: Effect of scale on the complexity of snowmelt systems. *Nordic Hydrology*, 12: 225–34.

Campbell, I. A. 1992: Spatial and temporal variations in erosion and sediment yield. *Erosion and Sediment Monitoring Programmes in River Basins*, International Association of Hydrological Sciences Publication 210: 455–65.

Cambers, G. 1976: Temporal scales in coastal erosion systems. *Transactions of the Institute of British Geographers*, ns1: 246–56.

Chappell, A., Oliver, M. A., Warren, A., Agnew, C. T., Charlton, M. 1996: Examining the factors controlling the spatial scale of variation in soil redistribution processes from southwest Niger. In M. G. Anderson and S. M. Brooks (eds), *Advances in Hillslope Processes*, vol. 1. Chichester: John Wiley, 429–49.

Church, M. 1996: Space, time, and the mountain – how do we order what we see? In B. L. Rhoads, and C. E. Thorn (eds), *The Scientific Nature of Geomorphology*, New York: John Wiley, 147–70.

DeBoer, D. H. 1992: Hierarchies and spatial scale in process geomorphology: A review. *Geomorphology*, 4: 303–18.

Douglas, I. 1988: Restrictions on hillslope modelling. In M. G. Anderson (ed.), *Modelling Geomorphological Systems*. New York: John Wiley, 401–20.

Fan, Y., Bras, R. L. 1995: On the concept of a representative elementary area in catchment runoff. *Hydrological Processes*, 9: 821–32.

Friedman, J. M., Osterkamp, W. R., Scott, M. L., Auble, G. T. 1998: Downstream effects of dams on channel geometry and bottomland vegetation: Regional differences in the Great Plains. *Wetlands*, 18: 619–33.

Haigh, M. J. 1987: The holon: Hierarchy theory and landscape research. *Catena* suppl. 10: 181–92.

Harrison, S. 1999: The problem with landscape. *Geography*, 84: 355–63.

Huggett, R. J. 1995: *Geoecology: An Evolutionary Approach*. London: Routledge.

Imeson, A. C., Lavee, H. 1998: Soil erosion and climate change: The transect approach and the influence of scale. *Geomorphology*, 23: 219–27.

Johnson, D. L., Keller, E. A., Rockwell, T. K. 1990: Dynamic pedogenesis: New views on some key soil concepts, and a model for interpreting Quaternary soils. *Quaternary Research, 33*: 306–19.

Juillard, E. 1962: The region: An essay of definition. *Annales de Geographie*, 71: 429–50.

Larson, M., Kraus, N. C. 1995: Prediction of cross-shore sediment transport at different spatial and temporal scales. *Marine Geology*, 126: 111–27.

Magilligan, F. J., Phillips, J. D., James, L. A., and Gomez, B. 1998: Geomorphic and seidmentological controls on the effectiveness of an extreme flood. *Journal of Geology*, 106: 87–95.

Martin, P. 1993: Vegetation responses and feedbacks to climate: A review of models and processes. *Climate Dynamics*, 8: 201–10.

Nicholson, S. 2000: Land surface processes and Sahel climate. *Reviews of Geophysics*, 38: 117–39.

O'Neill, R. V. 1988: Hierarchy theory and global change. In T. Rosswall, R. G. Woodmansee, and R. G. Risser (eds), *Scales and Global Change*. New York: John Wiley, 29–45.

O'Neill, R. V., DeAngelis, D. L., Waide, J. B., Allen, T. F. H. 1986: *A Hierarchical Concept of Ecosystems*. Princeton, NJ: Princeton University Press.

Pahl-Wostl, C. 1995: *The Dynamic Nature of Ecosystems. Chaos and Order Entwined*. Chichester: John Wiley.

Penning-Roswell, E., Townsend, J. R. G. 1978: The influence of scale on factors affecting stream channel slope. *Transactions of the Institute of British Geographers*, ns3: 395–415.

Pereira, V., Fitzpatrick, E. A. 1998: Three-dimensional representation of tubular horizons in sandy soils. *Geoderma*, 81: 295–303.

Phillips, J. D. 1986a: Sediment storage, sediment yield, and time scales in landscape denudation studies. *Geographical Analysis*, 18: 161–7.

Phillips, J. D. 1986b: Spatial analysis of shoreline erosion, Delaware Bay, New Jersey. *Annals of the Association of American Geographers*, 76: 50–62.

Phillips, J. D. 1988: The role of spatial scale in geomorphic systems. *Geographical Analysis*, 20: 359–68.

Phillips, J. D. 1995: Biogeomorphology and landscape evolution: The problem of scale. *Geomorphology*, 13: 337–47. Also published in *Biogeomorphology: Terrestrial and Freshwater Systems* (Proceedings of the 26th Binghamton Geomorphology Symposium).

Phillips, J. D., Perry, D., Carey, K., Garbee, A. R., Stein, D., Morde, M. B., Sheehy, J. 1996: Deterministic uncertainty and complex pedogenesis in some Pleistocene dune soils. *Geoderma*, 73: 147–64.

Phillips, J. D. 1997a: Humans as geological agents and the question of scale. *American Journal of Science*, 297: 98–115.

Phillips, J. D. 1997b: Simplexity and the reinvention of equifinality. *Geographical Analysis*, 29: 1–15.

Phillips, J. D. 1999a. *Earth Surface Systems. Complexity, Order, and Scale*. Oxford, UK: Blackwell Publishers.

Phillips, J. D. 1999b: Spatial analysis in physical geography and the challenge of deterministic uncertainty. *Geographical Analysis*, 31: 359–32.

Phillips, J. D. 2001: Sedimentation in bottomland hardwoods downstream of an east Texas dam. *Environmental Geology*, 40: 860–68.

Poesen, J. W., Torri, D., Bunte, K. 1994: Effects of rock fragments on soil erosion by water at different spatial scales: A review. *Catena*, 23: 141–66.

Reed, R. A., Peet, R. K., Palmer, H. W., White, P. S. 1993: Scale dependence of vegetation environment correlations: a case study of a North Carolina piedmont woodland. *Journal of Vegetation Science* 4: 329–40.

Ritsema, C. J., Dekker, L. W., Nieber, J. L., Steenhuis, T. S. 1998: Modeling and field evidence of finger formation and finger recurrence in a water repellent sandy soil. *Water Resources Research*, 34: 555–67.

Rosenstein, M. T., Collins, J. J., DeLuca, C. J. 1993: A practical method for calculating largest Lyapunov exponents from small data sets. *Physica D*, 65: 117–34.

Schaffer, W. M. 1981: Ecological abstraction: The consequences of reduced dimensionality in ecological models. *Ecological Monographs, 51:* 383–401.

Schumm, S. A., Lichty, R. W. 1965: Time, space, and causality in geomorphology. *American Journal of Science*, 263: 110–19.

Seyfried, M. S., Wilcox, B. P. 1995: Scale and the nature of spatial variability: field examples having implications for hydrologic modeling. *Water Resources Research*, 31: 173–84.

Sherman, D. J. 1995: Problems of scale in the modeling and interpretation of coastal dunes. *Marine Geology*, 124: 339–49.

Smit, A. 1999: The impact of grazing on spatial variability of humus profile properties in a grass-enroached Scots pine ecosystem. *Catena*, 36: 85–98.

Turner, M. G., Romme, W. H., Gardner, R. H., O'Neill, R. V., Kratz, T. K. 1993: A revised concept of landscape equilibrium: disturbance and stability on scaled landscapes. *Landscape Ecology*, 8: 213–27.

Twidale, C. R. 1998: Antiquity of landforms: an "extremely unlikely" concept vindicated. *Australian Journal of Earth Sciences*, 45: 657–78.

Williams, G. P., Wolman, M. G. 1984: Downstream effects of dams on alluvial rivers. *U.S. Geological Survey Professional Paper*, 1286: 1–77.

5 Embedded Scales in Biogeography

Susy S. Ziegler, Gary M. Pereira, and Dwight A. Brown

At the simplest level, biogeographical research includes three steps: observation, analysis, and representation. Most representations are in map form, but when the observational unit is the whole earth or a single point, representation is commonly a graph that focuses on differences in a phenomenon over time. When point observations are collected in sufficient spatial density, they can be mapped.

Biogeographic information is necessarily analyzed with other environmental traits (e.g., soil characteristics, amount of carbon storage in biomass, pollution levels) to address concerns about the effects of human impact, changing climate, and a broad array of global changes on the geography of organisms. It is necessary to understand how scales are embedded in the resultant biogeographical data or maps in order to avoid making improper assumptions about scale or imbuing the results with unwarranted confidence. Scale issues vary with the type of biogeographical research, depending on what the output is and whether time is a factor. It should further be noted that cartographic scales don't readily apply to Geographic Information Systems (GIS) – only to their map output. GIS analysis is constrained rather by the intrinsic uncertainty of data. We contend that the data with the greatest uncertainty establish the overall level of uncertainty.

Maps are the lynch pins of biogeography. They relate observations of biota to space and enable analyses of spatial patterns. Maps of biota come in many cartographic scales, but the idea of scale is more than just a characteristic of the representation of earth phenomena on a map. External factors such as printing limitations often dictate the map representation scale or cartographic scale, but the final cartographic scale often requires adjustment of many other types of scale. In other words, the multiple scales of biogeography are nested, and a radical change in one scale element requires changes in others.

We begin the chapter with brief discussions of scale issues related to biogeography and how observations are abstracted when data are generalized. This chapter then explores the way multiple scales have been embedded in maps of plant assemblages, maps of plant taxa and change over time, and maps of biogeographical processes. We draw our examples from a variety of disciplines that contribute to the biogeographical literature.

Scale Issues in Biogeography

The tendency of many discussions of scale is either toward imprecision or nondefinition, or toward an articulation of the differences among the elements of grain, extent, time and space (Pereira, 2002a). The more precise discussions often include a table of scale types with its roots in one produced by Delcourt et al. (1982; Schoomaker, 1998; Turner et al., 2001). Phillips (1995) distinguished among four elements of time for biotic disturbances, increasing the level of detail of the temporal scale. The specific time scales are the rate at which processes work, the frequency of occurrence, the duration of disturbance events, and the length of time between events for vegetation to respond. We shall not examine these in detail, except to note that they play a part of the embedded complexity in the fire disturbance cases we examine later.

Because of the complexity of the biosphere and its multitude of elements and functions, biogeography introduces an additional complex element of scale – that of attribute scale. Some attribute grain scales are simple linear measurements such as fractions of millimeters for tree-ring widths or centimeters in stem diameter. For entities that are classes like species or assemblages, the classifications generally minimize dissimilarity within a class and maximize the values between classes. Thus, species or plant assemblage variability has a scale grain. The scale extent of tree rings is the maximum age, for stem diameter it is the maximum size, and for species or assemblages it is the number of classes considered.

It is not our intent to produce a rigid nomenclature of scale types; we leave that to others who are more inclined to do so (Allen, 1998). Rather we want to illustrate the complexity of scale in biogeography. We do this through examples of a variety of maps and other products of biogeographical analysis. We attempt to define our terms carefully, with the understanding that they may have been defined differently by others. We use a variety of examples to illustrate how scales are embedded in biogeography. Those examples that do not involve a map represent a geographic unit that ranges from a point to the entire earth.

We concern ourselves in this chapter mostly with maps of biota, biotic processes, and biotic disturbances. Such maps are produced in a wide array of disciplines. The analyses range from reconstruction of fire histories from fire scars recorded in tree rings for the Boundary Waters Canoe Area Wilderness (Heinselman, 1973) to dispersal of disease organisms by the Centers for Disease Control. The map subject data come from myriad sources including field observation of individual trees, satellite monitoring of individual caribou or duck migrations, and remote sensing of heterogeneous pixels ranging in size from one to several thousand meters. The maps may show nominal occurrence, classifications of plant assemblages, population density, quantity of biomass, mass transfer, or may be representations of a time series that focuses on assessment of change. To understand some of the embedded scales, it is necessary to delve into some debates in ecological theory to tease apart the observation scales and how they have been transformed in the cartographic process.

Observation imposes the ultimate limit on all elements of scale. It is similar to the least significant figures rule in mathematics and its corollary in geographic analysis that the variable with the coarsest spatial grain dictates the grain of the result. Understanding the meaning of a biotic map requires that the embedded scales and all transformations are specified or can be correctly inferred. Unfortunately, not all biotic maps are accompanied by full descriptions of the relevant information on the observations or the transformations of attributes and space that created the final map.

Observations start with classification, and often proceed through subsequent steps of classifications. Even if data are collected in a simple tally, the tally counts things belonging to discreet classes. For instance, measurement of carbon flux in mg/m^2/second is done in equal range classes that are defined by the minimum instrument precision. Ground-based observations are often made at the species level. Data may be abstracted into populations as in the counting of birds of a given species, or further into relative abundance as in the change in the populations of two tree species after a fire, or into assemblage classes such as defining multiple species observations as "oak-hickory forest." For remote sensing observations, the raster cell is the fundamental unit that must be characterized by interpreting spectral signatures into a meaningful class of phenomena, such as a forest type.

Abstraction of Observations

Abstraction of observations results from the generalizations that often accompany many of the steps in biogeographical research. We can see the product of these generalizations only by dissecting the resultant map or graph to see what scale elements are involved. The complexity of scale elements applied to mapping biota involves up to nine scale elements (Table 5.1). Each element has its own measurement units. Not all mapped data are real numbers; some are ordinal, and others are the zero to one bounding of probability maps. In some cases the units may be merely implied in the selected categories such as the dissimilarity of values or traits imbedded in taxonomic levels. For example, there is more similarity of included organisms at the species level than at the genus or family levels. Mere presence or absence has an attribute granularity of one.

The anatomy of embedded scales varies substantially by subject. Table 5.2 shows some examples of embedded scales in a variety of biogeographical works. They are arranged by studies of an organism or assemblage of species, and studies that examine a variety of processes that change biotic patterns. In some cases, we provide graphic examples and in others we give verbal descriptions. The attribute scale is complex in that it may involve multiple attributes on a single map and the scale units are varied number counts, measures of magnitude, or specified or unspecified dissimilarity (Table 5.2).

Table 5.1. *Embedded scale types in biogeography*

Scale element	*Characteristic units*
Observation Space Grain (unit diameter)	cm, km
Observation Space Extent	cm^2, km^2
Attribute Type Grain	Numerical or implied dissimilarity index
Attribute Quantity Grain	Population count, percent, ordinal scale, etc.
Attribute Extent	Range of attribute values
Time Grain	Seconds, millennia
Time Extent	Days, eons
Cartographic Scale	Very small to large (representative fraction)
Cartographic Grain	May be different from observation grain; in earth units it remains constant with enlargement

Table 5.2

Biogeography scale		Observation						Map scale		Ref.
		Space		Attribute	Time			Abstractions		
Tab/Fig	Examples	Grain diam. (m)	Extent diam. (km)	Grain, n or unit*	Grain (yr)	Extent (yr)		Grain (km)	Scale 1:	
	Organism/assemblage									
F 5.2a	*A. gerardii* landscape abund.	0.01	5	1	1	1		0.10	100,000	Brown & Brown
F 5.2b	*A. gerardii* abundance	50	2,500	1	30	1		50	10,000,000	Brown
F 5.2c	*Andropogon* diversity	500	5,000	50	150	1		400	20,000,000	Brown
F 5.2d	Andropogoneae importance	1,000	12,200	200	150	1		500	150,000,000	Brown
F 5.3	Oaks species probability	50	1,000	1	3	1		50	15,000,000	Brown
F 5.4b	Forest density, US	1	5,000	5	0.01	1		10	20,000,000	USDA -FS
	Biomes	100	12,200	1,000	100	1		500	150,000,000	Küchler/Koppen
F 5.1	Potential Natural Veg., US	0.3–100k	5,000	100	NA	NA		10	5 to 14mil.	Küchler '64
F 5.1	Natural Vegetation, MN	0.3	500	80	NA	NA		5	500,000	Marschner
F 5.1	VEMAP Potential Nat. Veg.	0.3–100k	5,000	600	NA	NA		50	20,000,000	VEMAP
	Seasonal Land Cover, US	1000	5,000	80	NA	NA		1	7,500,000	Loveland et al.
	Stand structure versus age	1	100	<1 mm	1	88–390		NA	NA	Ziegler
T 5.3	FIA database	0.1	5,000	1	1	3		NA	NA	USDA -FS
	Diversity change									
F 5.5	Global diversity change	50,000	12,200	1,000	1,000,000	2 billion		NA	NA	Benton
F 5.6	Local pollen change	0.10	1	5	250	30,000		NA	NA	Fredlund
	Mammal pop. loss Australia	1,112,000	2731	6, 4, 25	variable	variable		1,112	158,986,200	Ceballos & Ehrlich

(*continued*)

Table 5.2 *(continued)*

Biogeography scale		*Observation*					*Map scale*		*Ref.*
		Space		*Attribute*	*Time*		*Abstractions*		
Tab/Fig	*Examples*	*Grain diam. (m)*	*Extent diam. (km)*	*Grain, n or unit**	*Grain (yr)*	*Extent (yr)*	*Grain (km)*	*Scale 1:*	
	Mammal pop. loss Africa	1,112,000	5538	5, 4, 25	variable	variable	1,112	317,972,350	Ceballos & Ehrlich
	Mammal pop. loss N. Am.	1,112,000	4972	3, 2, 25	variable	variable	1,112	370,967,750	Ceballos & Ehrlich
	Dispersion								
F 5.10	Caribou migration	0.10	440	1	0.01	1	1.2	5,529,400	Thomas/NBS
F 5.7	Chestnut spread	0.10	1,000	1	250	15,000	50	20,000,000	Davis
F 5.8	Chestnut blight spread	0.10	1,000	1	10	50	50	20,000,000	USDA-FS
	Disturbance								
	Fire detection (global)	1,000	12,200	1	0.25	1	fine	380,000,000	Fuller et al.
F 5.4	Forest change (US)	10	5,000	500	1	300	10	20,000,000	USDA-FS
	Fire burn extent (Indonesia)	1,000	800	1	0.02	1	fine	10,000,000	Dwyer et al.
F 5.11	Fire history (BWCA)	1	115	1	1	361	2	1,280,000	Heinselman
	Fire modeling (Amazon)	1–1,000	10–100	1	.003	.01	.001	100,000	Pereira
	Production, growth								
F 5.9	Biomass prod. (SeaWiFS)	1,000	22,457	100	1	3	20	150,000,000	NASA/GSFC
	Tree growth (tree rings)	1	.01–500	<1mm	1	95	100	50,000	Cook et al.

*Attributes may be multiple for one map. Grains may be number, magnitude, or specified or unspecified dissimilarity.

Embedded Scales in Maps of Plants Assemblages and Land Cover

Plant assemblage maps

The oldest and coarsest observational, attribute and spatial scales are found in the *klima* (*clima*) concept of early Greek and Latin scholars that divided the earth into latitudinal bands into zones of similar environments with such descriptions as frigid, temperate, and torrid. In this concept, biota is firmly linked to the sun angle and therefore climate. At the dawn of the twentieth century, Frederic E. Clements added considerable detail to the climatically determined patterning of biota with his climax concept. His top down system provided a classification into which field observations could be grouped. The strong connection between biota and climate, and the stability of climate, were the unquestioned foundations of his system.

Although Clements failed to map his climax model, other people did. Indeed, his ideas survive today because they are matched with a simple cartographic paradigm of the dasymmetric map (area symbolization with classed polygons). The most general example is Vladimir Köppen's (1931) climate map that used vegetation to fill in climate boundaries where climate data were sparse or nonexistent. We hypothesize that biome maps, which are common in geography and ecology textbooks, developed from Köppen's climate map.

Nesting of spatial scales

Subsequent maps of natural vegetation were strongly influenced by Clements' idea, but it is not easy to determine the observational and classification basis of these maps because the cartographers failed to explain how the classes were derived or how map boundaries were determined. The United States map of "Natural Vegetation" by Homer L. Shantz and Raphael Zon (1924) printed in the *Atlas of American Agriculture* is the first example. Although Shantz and Zon did not explain how they created their map, there is a logical connection between their map and Clements' concepts, since Zon and Clements probably had contact at the University of Minnesota. Nicholson (1990: 141) describes N. (sic) L. Shantz as an adherent to the "community-unit theory" of Clements which is highly probable because both were early Ph.D. advisees of Charles Bessey at The University of Nebraska. Thus, we infer that Shantz and Zon classified observations into climax-vegetation categories. There is no doubt that field observations tempered the application of Clements' top down system.

A. W. Küchler's (1964) map titled *Potential Natural Vegetation of the Coterminous United States* is cut from the same cloth. Küchler's first map drew heavily upon the Shantz and Zon map (Küchler 1949). His second version was published in 1964 and 1966 at a slightly smaller scale in *The National Atlas*. A sample of Küchler's map is shown in Figure 5.1b. We chose to show this area of his map because we could compare it to a manuscript

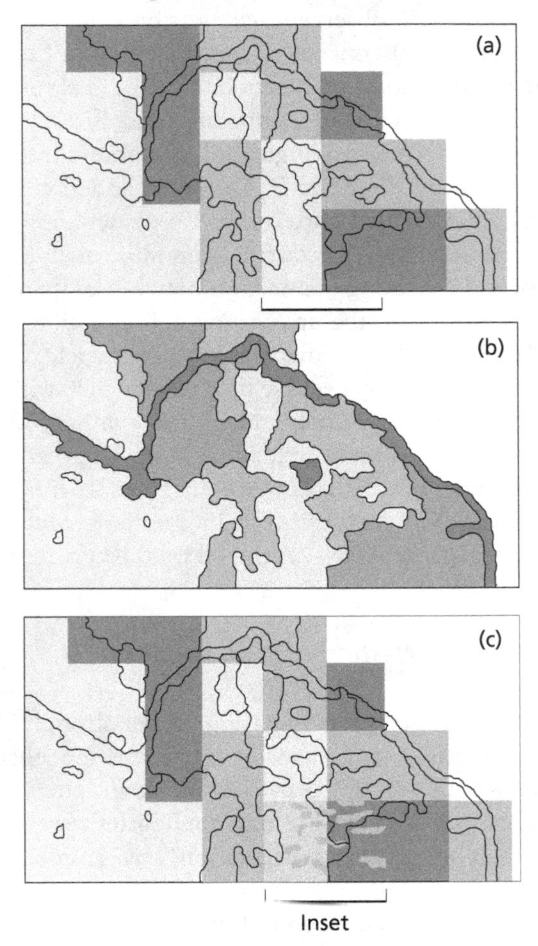

Figure 5.1 *Scale and attribute generalization of vegetation in Southeastern Minnesota. Different grain scale renditions of vegetation cover in Southeastern Minnesota: (a) is the VEMAP project gridded rendition of the Küchler map with an overlay of Küchler's boundaries. It includes one prairie and two forest classes; (b) is Küchler's map with one prairie and three forest classes; (c) has an inset from Marschner's map over both the gridded VEMAP version and Küchler's boundaries. The inset from Marschner's map included two prairie and five forest vegetation classes in the small inset area. The two VEMAP cells in the inset include one prairie and one forest class.*

map by Francis Marschner that Küchler (based on the similarity of major boundaries) had undoubtedly seen. Marschner based his *Original Vegetation of Minnesota* map on Public Land Survey surveyor notes. He abstracted the potential eight observations per section, for a total of over 650,000 sites, onto a map that used a minimum mapping unit of about 15 square km. A one-degree longitude by half a degree latitude inset is shown in Figure 5.1c. Küchler's map includes one prairie and three forest classes for this area. Marschner's map grouped individual tree species and plant cover observations into wet and dry prairie and five forest classes for the same area. Thus Küchler increased the attribute granularity of this area.

H. A. Gleason, a contemporary of Clements, Shantz, and Zon, found little evidence in his extensive field experience to support Clements' views of the structure of climax assemblages, and he concluded along with numerous subsequent biogeographers and ecologists that Clements' climatic determinism approach was seriously wanting. Gleason advanced a nondeterministic view of the dynamics and patterning of plants that he termed "the individualistic concept." His model would fit into what are now referred to as "complex systems models." These models represent conceptual frameworks for understanding biota of the earth that are only enabled by increasing the detail (decreasing grain size) of our observations. Figures 5.2, 5.3, 5.7, and 5.8 draw heavily on Gleason's model of vegetation change, which involved chance, proximity, and dispersal.

O'Neil (1998: 5) argued that "considerations of scale provide a reconciliation of Clements' (1916) continental picture of plant communities constrained by climate and Gleason's (1926) analysis of individual plant communities, where competition is the critical process." He misses the point of both ecologists, who were describing different views of local processes of change. Clements' climax vegetation is easily mapped, while Gleason's individualistic model is very difficult to map because the attribute grain scale is much finer and the attributes of space are dynamic as plant assemblages change over time.

Recent deterministic, climate-based models used to predict future climax vegetation patterns are a reversion to Clementsian concepts (VEMAP members 1995; Pan et al., 1998; Schimel et al., 2000). The VEMAP project used Küchler's map as the base from which to model changes and greatly increased the attribute and spatial granularity of his map. The VEMAP members used the climate output of various global climate models and scenarios to predict how the Küchler map would change when run with a variety of vegetation simulation models. Their starting base for southeastern Minnesota abstracts Küchler's four assemblage classes into three (Figures 5.1a and 5.1c). It lumps seven Marschner map classes (22,050 corner observations) into two classes. We can compare these generalized maps to a sample of recently observed trees from the FIA database (Table 5.3). The Marschner map shares an observational base with the FIA data, and the Küchler map shares this base only indirectly through adoption of Marschner's boundaries and class descriptions. Marschner names more taxa in his legend than does Küchler. The single VEMAP class corresponding with the right half of the inset in Figure 5.1c captures taxa that make up only 17 percent of the composition of the FIA (Federal Inventory and Analysis) sample.

In the above cases, as in many other examples, vegetation types traditionally are mapped at the nominal level because it is impossible to collect

Table 5.3 *Attribute scale comparisons*

	FIA data for 1988		*Taxa presence implied by forest class descriptions in 1° by 1/2° inset in Figure 5.1c*		
Native species (20)	*Count*	*Percent*	*Marschner*	*Küchler*	*VEMAP*
Eastern white pine	4	0.85%	X		
Boxelder	29	6.14%			
Silver maple	5	1.06%	X	X	X
Sugar maple	26	5.51%	X	X	X
Bitternut hickory	3	0.64%	X		
Hackberry	5	1.06%			
White ash	6	1.27%	X	X	
Green ash	18	3.81%	X	X	
Butternut	1	0.21%	X		
Black walnut	7	1.48%	X		
Eastern hophornbeam	10	2.12%			
Quaking aspen	1	0.21%	X	X	
Black cherry	6	1.27%	X		
White oak	31	6.57%	X	X	
Bur oak	100	21.19%	X	X	
Northern red oak	105	22.25%	X	X	
Black oak	7	1.48%	X	X	
American basswood	47	9.96%	X	X	X
American elm	42	8.90%	X	X	
Slippery elm	13	2.75%	X	X	
Recent arrivals (3)					
Eastern red cedar	2	0.42%			
Scotch pine	3	0.64%			
Choke cherry	1	0.21%			
Total (23) / % of FIA	472	100.00%	89%	85%	17%

information on their continuous geographic variation. We often depend on these historic maps to give context to maps of predicted change. Meanwhile, change studies are hampered by nonstandard attributes, such as classes comprised of different species groupings. The above three maps have a common origin and thus the classes are crudely nested, with the most detailed map serving as the basis for subsequent maps. Unlike the case above where the spatial scales and attribute scales are nested, Franklin and Woodcock (1997) found that for two vegetation databases in southern California, the data were spatially nested, but not taxonomically nested.

North American seasonal land-cover maps

Remote sensing provides a tool with a finer spatial grain than was available for production of any of the previous examples except for the observational grain of the individual trees in the FIA database or pollen grains in pollen analysis. However, not all biotic maps based on remotely sensed data are presented at the observational spatial grain (resolution). For a continuous index like the Normalized Difference Vegetation Index (NDVI), the classification of NDVI maps into successively larger cells is a straightforward process of generalization with a corresponding loss of information (De Cola, 1997). This generaliztion process is not at all like changing the spatial grain of nominal classifications of polygons (i.e., merging category descriptions), which requires a judgment about dominance or some other subjectively chosen importance criterion.

Loveland et al. (1995) and others have used the seasonal change of spectral signatures of Advanced Very High Resolution Radiometer (AVHRR) – remote sensing images to produce a land-cover map of the United States with 154 cover classes (Table 5.2). The initial spatial resolution (observational grain) is 1,000 meters, and the cartographic grain is commonly held the same. The observed attribute grain of the earth's reflected or emitted energy measured by the satellite varies by wavelength band on the NOAA-11 satellite sensor used to collect AVHRR data. These energy measurements are divided into scales with 1,024 to 256 steps depending on the wavelength of the sensed band of the electromagnetic spectrum. The Loveland et al. land-cover map includes both natural and managed vegetation classes. It has nearly twice as many classes as Küchler's map. In other words, there is less within class dissimilarity, and hence a finer attribute grain.

Embedded Scales in Maps of Plant Taxa and Taxa Change

Mapping plant taxa

Studies in ecological biogeography commonly focus on individual plants or clonal clumps within quadrats. The data collected from plots can be pooled to study the spatial extent of a plant species. Beyond the range of a species, we have to relax the attribute taxonomic scale or attribute grain to include near relatives such as those plants of the same genus. In so doing, we expand the dissimilarity of the subjects under consideration. To look at global relationships of plants often requires that we further expand our dissimilarity tolerance to find connections with even more distant relatives. For instance, the grass family, because of its extraordinary geographic range, provides some good examples of attribute scale changes forced by a change in geographic extent from landscape to regional to national to global (Figure 5.2).

In the big bluestem (*Andropogon gerardii*) example (Figure 5.2) the available data constrain the attribute detail (Brown, 2002). At the local, landscape, and regional observational scales, quantitative data are available. As we expand the extent of our observations, the only data we have a reasonable chance of finding are species lists, whereby we can only talk about the presence or absence of a particular taxon in a large area. Beyond national boundaries, the uniformity of public data – or their mere existence – is less likely. In the case of big bluestem, regional and national scale portrayals of abundance reflect the geographic pattern of dominance for the Andropogoneae tribe, of which big bluestem is a member.

Similar detailed data are available for trees from sample plots in the FIA database of the United States Forest Service. Numerous characteristics of these individual trees are recorded and linked to plots that are further described by attributes including location, site hydrological conditions, ownership type, stand age, and other tree and site variables. Figure 5.3 shows the probabilities that a live, large tree in a stand originating before 1901 is one of three dominant oak species. The probabilities range from zero to one and are calculated at the county level in order to have a large enough sample size.

The attribute of "chance of occurrence", as in the oaks example, is a form of probabilistic and fuzzy logic applied to cartographic efforts (Figure 5.3). What are the probabilities that large trees in old stands belong to a class, in this case a certain tree species? Three species are shown in Figure 5.3, but any such map implies at least one additional class. If only one species is included, for example bur oak, the classes would be bur oak and not bur

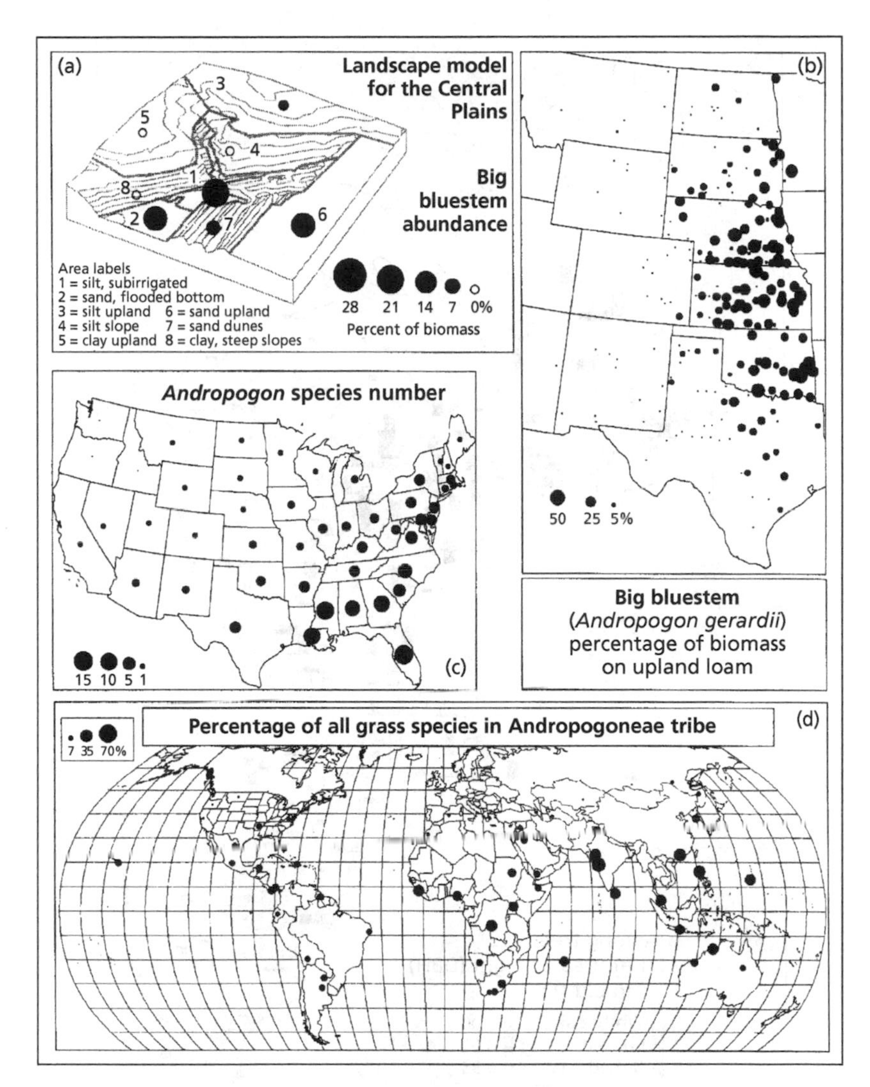

Figure 5.2 *Multiple scale representations of big bluestem (*Andropogon gerardii*) relations. Smallest dots in b and d indicate none observed.*

Source: (a) is modified after Brown and Brown, (1996;) (b), (c), and (d) are from Brown, (2002).

oak. The probabilities of "not bur oak" are one minus the probabilities of bur oak.

By filtering the FIA data by live status, a minimum diameter at breast height (DBH) of 12", and old stand origin we reduce the number of trees from over 700,000 to fewer than 20,000. The analysis moves us from an

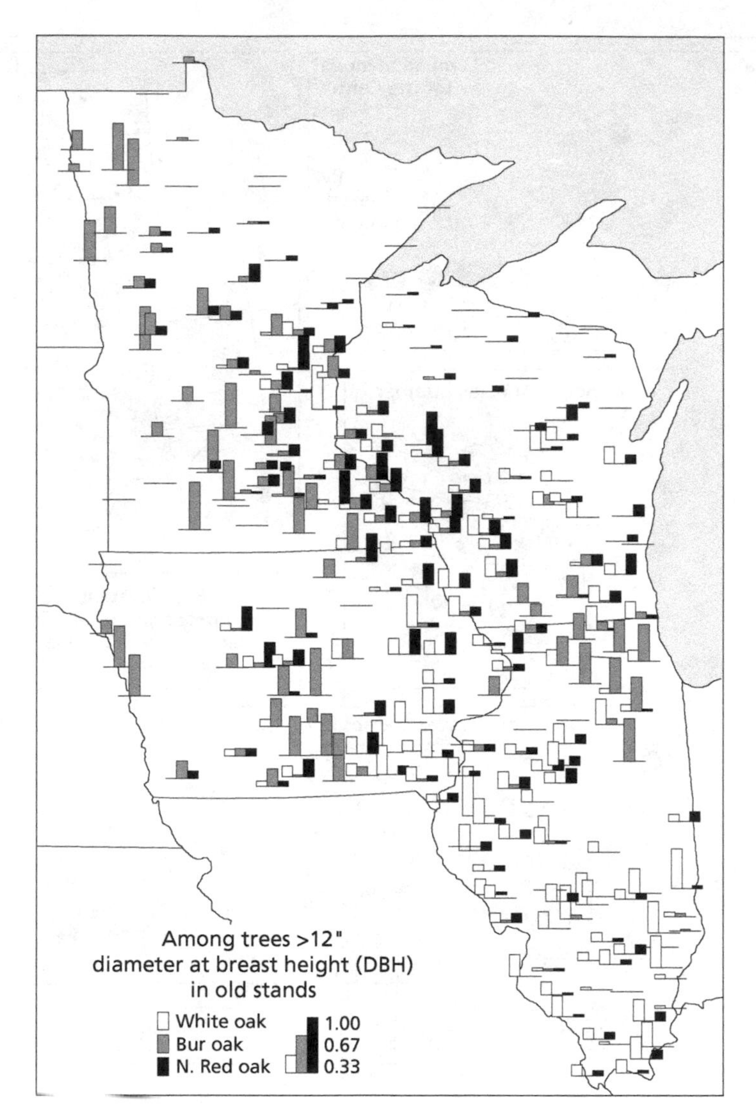

Figure 5.3 *Probabilities of three oak species among large trees in old stands.*

observation of a single tree assigned to a species class to sample populations by species for counties. In some counties the number of trees considered in this subset of data is more than one thousand, in other counties it is two orders of magnitude less. Thus the representative attribute grain varies with county. The least significant figures rule dictates that the grain for the whole map is 0.1, which is the significant figure for the county with the fewest qualifying trees. In this case the observational grain size is the individual, but the cartographic portrayal is the county. If the cartographic scale is changed, these same fundamental observational and attribute scales remain embedded in the map.

Taxa change and complex scale embedding

When trees are remotely sensed, the observational grain size is the resolution of the imaging system, the observational extent (i.e., geographical scale) is the coverage of the study area, and the attribute scale is defined by the algorithm used to convert a spectral signature into classes (Figure 5.4b). To study the extent of forest-cover change through time requires the use of historical information collected by different methods. The census of trees at one date may be simply an interpretation of some form of proxy data such as climate, soils, or topography. Thus some of the change over time derives from observing a phenomenon through different lenses (Figure 5.4a). The resultant map has an error band associated with the complex embedding of attribute scales.

A recent study of mammal population loss provides an example of complex scale embedding. Ceballos and Ehrlich (2002) compiled species range maps of different dates into a 2°-latitude by 2°-longitude grid of continents to examine range changes. It is rather straightforward to examine scales of their printed maps (Table 5.2). The complexity of embedding comes in the multiple transformations of the original field observations that lie behind the two generations of range maps. These start with records of occurrence that undoubtedly were observed at a sub-kilometer grain scale. For some species, such as the African elephant, the observations used to make the original range maps included Neolithic rock art, while other observations relied on historic records (Kingdon, 1997). Here, as in Asia where species on some maps are delimited with straight lines that span 30°-longitude, the observational grain of the source maps that is as high as a 5° by 5° grid is lost (Corbet and Hill, 1992).

The contemporary species range maps for Europe analyzed by Ceballos and Ehrlich (2002) used a 50-km grid that included the categories of documented presence and presumed presence (Mitchell-Jones et al., 1999). The grain of Mitchell-Jones et al.'s original maps is much finer than the grain of Ceballos and Ehrlich's maps, but the antecedent range maps do not reflect the grain of the initial observations.

The Ceballos and Ehrlich (2002) maps present variable map scales with a uniform grain size (a 2° by 2° grid) that is finer than some of the inputs and coarser than others. Using the least significant figures rule, the apparent grain size of a 2° by 2° grid should more realistically be a 5° by 5° grid. Thus

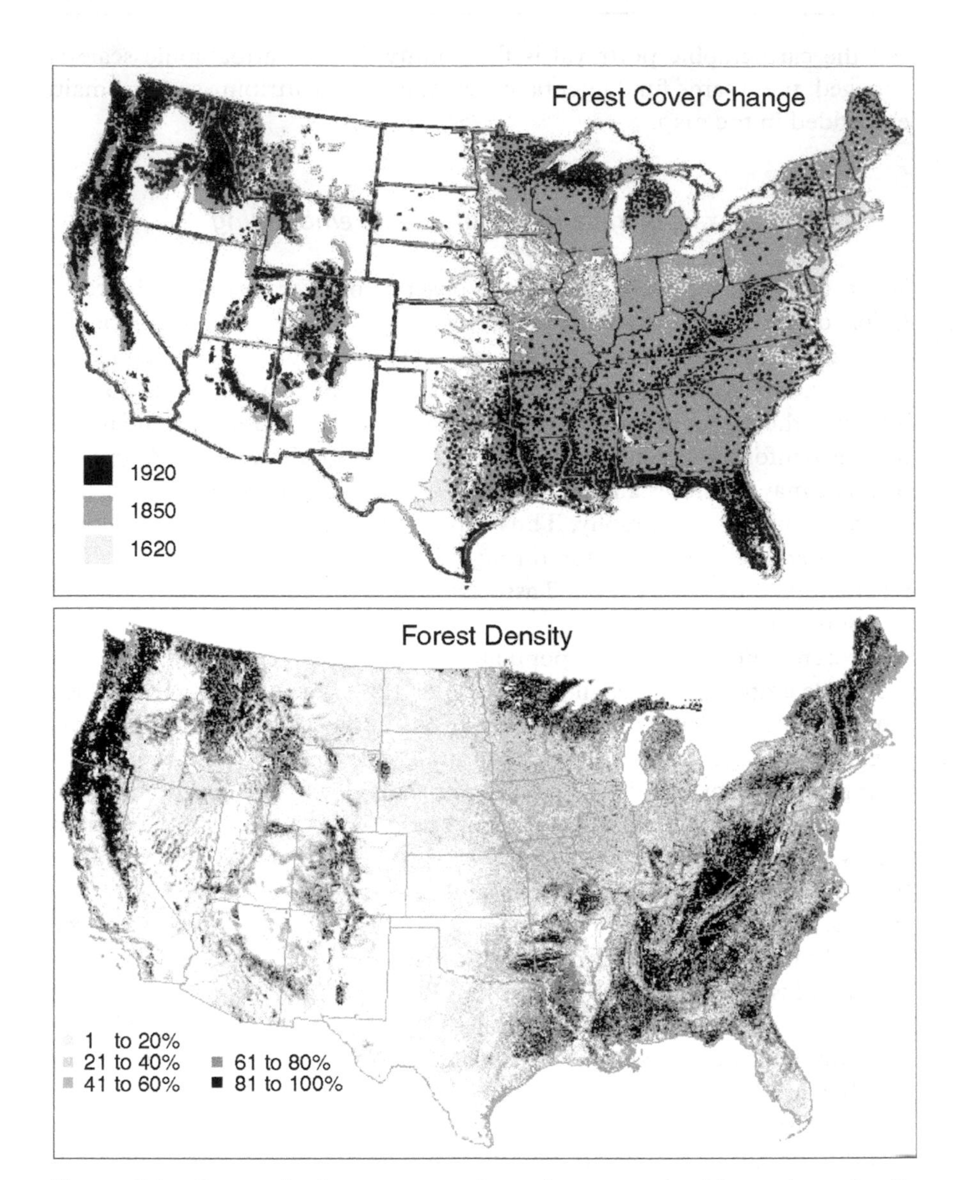

Figure 5.4 *Change in forest extent (upper) compared with modern density (lower) modified after USDA-Forest Service. Map (a) is based on multiple data sources; map (b) is based on remote sensing.*

Source: Modified from Grealey (1925) (map (a)), and the US Forest Service (map (b)).

the observation grain described for their map in Table 5.2 is at least two generations removed from the primary data that are already coarser grain transformations of the original observations.

Taxa change without cartographic scale

Not all biogeographical studies have a cartographic scale. Evolution and the history of increasing diversity of organisms on earth is one example (Figure 5.5). The spatial extent of observations is all of the earth, and the spatial grain is fine, but irregularly spaced. The scale of the time extent is similarly huge, but here the time grain, though variable, is coarse in millions to tens of millions of years. For the same reasons that tribe-level taxonomic data were needed to show global relations of big bluestem, here there is a need to define a taxonomic level (in this case families) at which fossils can be operationally defined with some precision. The observed attributes of individual fossils are abstracted to a broader, common category with coarse-grained similarity.

Pollen analysis of a single site provides the other extreme of observational extent. It represents a spatial sample of a circle (the core) about 5 cm in diameter. It is thought to represent an integrated sample of a lake of perhaps 100 m diameter or larger. The time extent is commonly a few thousand years because lakes and bogs that trap the pollen are very temporary features when seen in the extent of geologic time (Figure 5.6). The time extent of about 29,000 years in Figure 5.6 is long by pollen study standards. The time

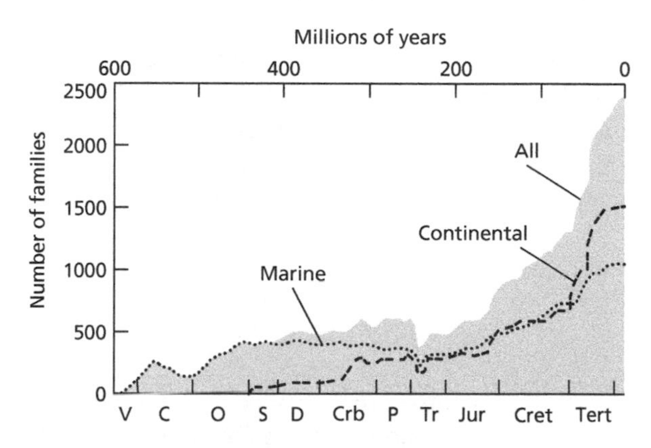

Figure 5.5 *Diversification record of the earth's biotic families since early Precambrian time (>600,000,000 yr BP). V=Vendian, C=Cambrian, O=Ordovician, S=Silurian, D=Devonian, Crb=Carboniferous, P=Permian, Tr=Triassic, Jur=Jurassic, Cret=Cretaceous, Tert=Tertiary.*

Source: Modified after Benton, (1995).

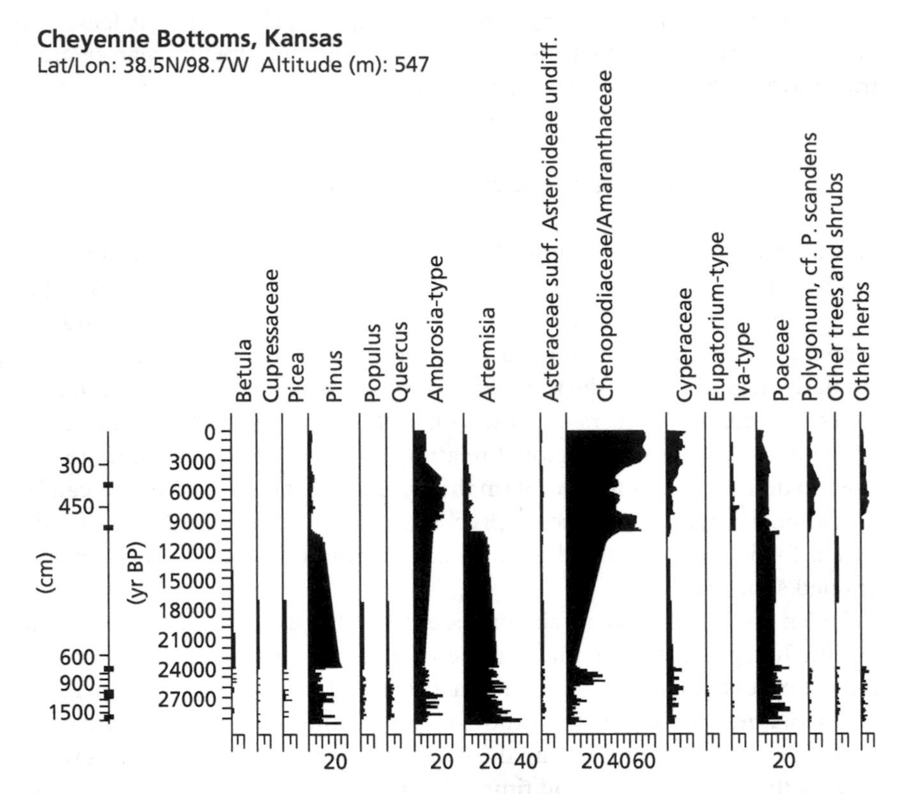

Figure 5.6 *Pollen record for Cheyenne Bottoms, Kansas.*
Source: Modified from Fredlund (1994).

resolution is variable because the core is sampled at uniform increments of depth. The middle record (i.e., the period from 24,000 to 9,000 yr BP) is sparsely sampled in time because sediment deposition was dramatically slower, and a thin layer spans a great deal of time. It might be inferred that landscape-scale disturbances were less frequent during this period of slow deposition.

Mapping taxa change

When numerous pollen records are compiled and mapped, they can present a spatial picture that has the same observational and attribute scales as the single cores, but the cartographic grain is determined by the density of control points used to make the map. Cartographic extent is limited by the intent of the study or by the extent of observational data. The cartographic scale and attribute scale are adjusted to meet the requirements of

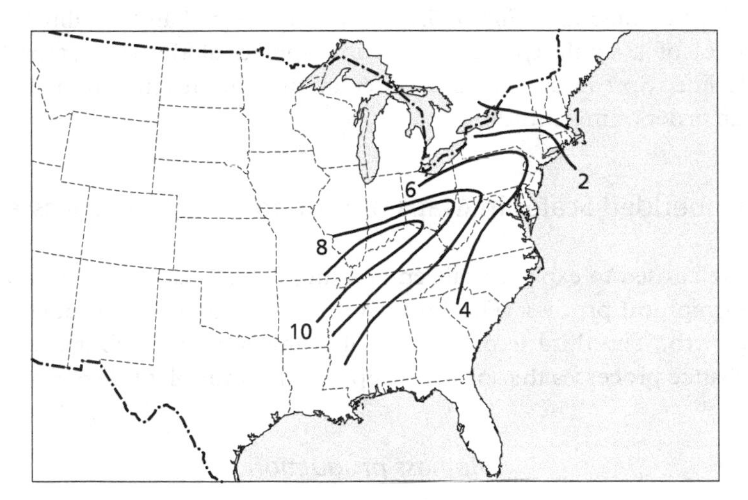

Figure 5.7 *American Chestnut spread. Isolines represent first arrival times of American Chestnut in 1000 years before present.*
Source: Modified after Davis (1983).

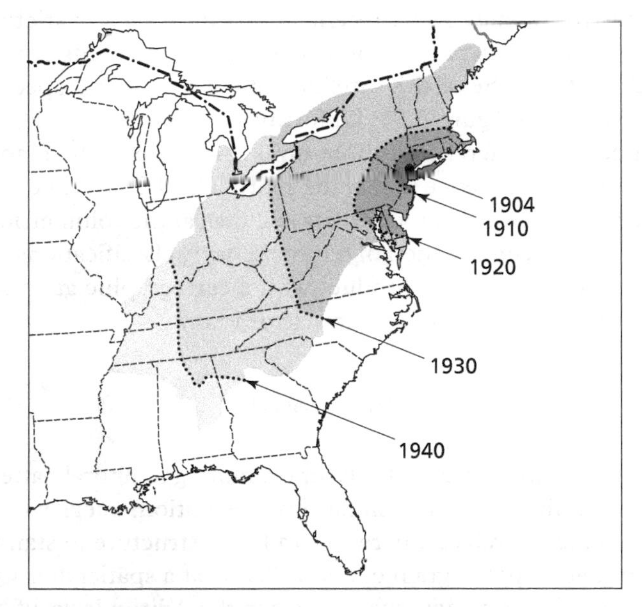

Figure 5.8 *American Chestnut blight spread. Chestnut blight was first discovered in 1904 at the Bronx Zoo in New York City. It is caused by a fungus named* Endothia parasitica *which forms a canker that enters a fissure in the bark and in a few years girdles the tree. It kills the above ground plant.*
Source: Modified after Flippo Gravatt (1949).

the published medium (Figure 5.7). Both Figures 5.7 and 5.8 illustrate the dynamics of a single species – first the spread of the species and then the decline. Spread and decline differ in time grain and extent by more than an order of magnitude.

Embedded Scales in Maps of Biogeographical Processes

We have chosen to explore scale embedding further using four examples of biogeographical process. The first two deal with biomass production and tree growth. The third is movement of animals, and the fourth concerns disturbance processes that provide opportunities for plant dispersal.

Biomass production

Attempts to portray the geographic differences in biotic functioning have been attempted at many scales from local to global. Quantifying global biomass production (net primary production) has been attempted by mapping the extent of biomes and assigning a characteristic annual production to the biomes. More recent studies have used a variety of remote sensing devices to assess the productivity of terrestrial plants or ocean productivity. AVHRR and SeaWiFS satellite data have been mapped to portray global productivity (Figure 5.9). Unfortunately the landmasses and water are presented with different attribute units and scales. The land areas are characterized by a difference index, the Normalized Difference Land Vegetation Index, and the ocean productivity is shown as the volumetric density of chlorophyll *a*. This combination of remote sensing classifications produces a very small cartographic scale product with a cartographic grain size of just over 30 times the observational grain size of 1 km.

Tree growth

Tree-ring records are attractive tools for examining temporal patterns of tree growth because they provide annual time resolution. Ziegler (2000) used dendrochronology to relate differences in forest structure to stand age. Her observations had a spatial grain of about 0.1 ha at a spatial density of about 0.5 km in each of three widely spaced sites in the Adirondacks of New York. The research did not result in a map, but graphs relate each structural characteristic to stand age as estimated from tree rings. Other observations included a census of the species composition of the canopy and understory by DBH, the stand basal area, and the volume of coarse woody debris. Each of

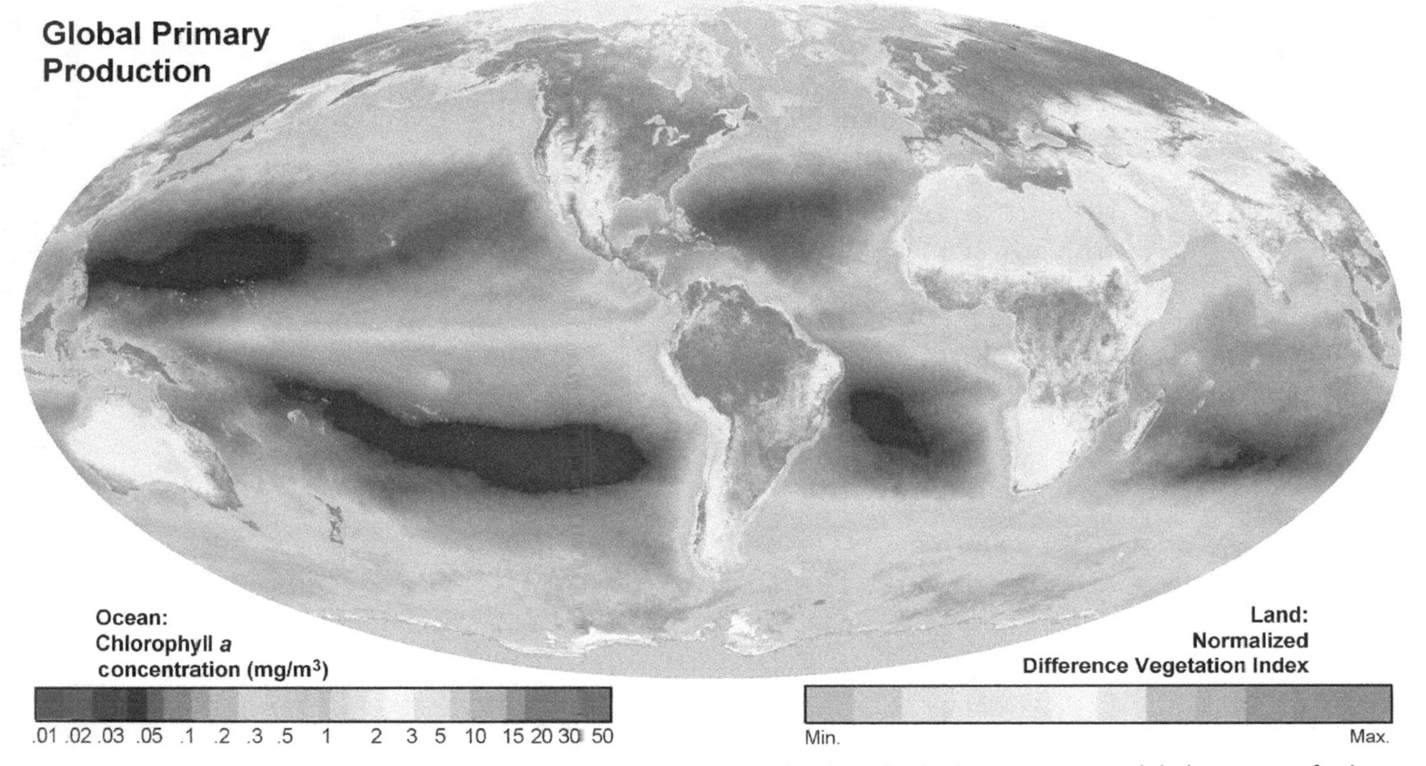

Figure 5.9 *Ocean chlorophyll and Normalized Difference Vegetation Index (NDVI). The image portrays global patterns of primary productivity from Sept. 1997 to Aug. 2000. Note the pattern of low ocean productivity over the subtropical oceans and the high productivity of coastal areas at high latitudes.*

Source: Image provided by ORBIMAGE. © Orbital Imaging Corporation and processing by NASA Goddard Space Flight Center. http://seawifs.gsfc.nasa.gov/seawifs.html.

these measurements and categorical classifications has its own grain, extent, within class diversity, and measurement or classification precision and error.

The annual time grain is difficult to find in other noninstrumented environmental records. Thus, there is pressure to apply dendrochronology information to a variety of purposes such as dating artifacts, and to seek a relationship with climate so that the tree-ring record may be used as a proxy for the instrumented climate record. Cook et al. (1992) used tree-ring records from over 150 sites in eastern North America to model the Palmer Drought Severity Index (PDSI). Over the period of record, tree rings explained about 45 percent of the drought index. While tree-ring widths are measured to a very fine tolerance (~0.01 mm) relative to the range of widths (ring widths may be up to three orders of magnitude greater than the measurement precision), the error band around the predicted PDSI is large. Thus, the attribute grain scale of the observed ring widths used by Cook et al. (1992) is finer than the predicted PDSI.

The appeal of an annual time grain also makes tree rings an attractive tool for reconstructing fire histories over large areas. The section on disturbance processes provides an example of scale embedding in the use of tree rings to reconstruct fire history.

Dispersal and movement

One of the most complex examples of embedded scales is the monitoring and mapping of caribou migration. This research involves a high level of attribute detail and high temporal resolution on a small-scale map (Figure 5.10). Each line represents movement of an individual over time. There is an extreme difference between observation grain of the transmitter (a few cm) and the map grain of about 1.2 km, but the attribute grain is the individual caribou in both cases. Both observational extent and cartographic extent are the same. All we know about this map, which disappeared without a trace from the Web, is that it represents a sample from a 17-year program of monitoring the movements of individual caribou, and that the tight, green clusters indicate the wintering grounds.

Disturbance processes

Fire is the most widely examined disturbance process, probably because it is easily observed by remote sensing methods, and it affects a wide range of biotic assemblages from the arctic to the tropical rainforests. The

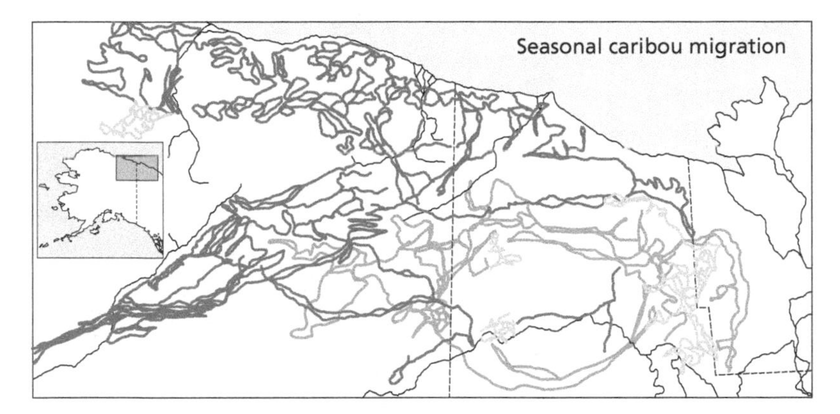

Figure 5.10 *Seasonal Caribou migrations in Arctic Slope. The light lines show movement around the wintering grounds; dark lines are seasonal movements.*
Source: Modified from Thomas (2001).

geographic extent of disturbance by fire may be determined from proxy data such as tree rings, from remotely sensed data, or through modeling.

A classic approach to understanding disturbance is the reconstruction of fire histories from fire scars recorded in tree rings, such as Heinselman (1973) completed for the Boundary Waters Canoe Area Wilderness. His series of maps used observations of tree location to interpolate fire extent for time periods of single years based on fire scars in tree rings. The extent of the mapped area is an irregular area about 75 by 180 km. While the observation is individual trees or stumps, the minimum cartographic grain is a fire extent of about two square kilometers. The time extent is 1610 to 1971, for which he mapped 93 polygons. The frequency and size of events varied throughout the record (Figure 5.11). There is not a sufficient distribution of 400-year-old trees to capture the extent of the oldest fires; thus, the functional observational extent increases through time. Not all temporal processes are amenable to the same type of analysis. While fire is an event with a disturbance rate or intensity, use of a duration scale similar to Phillips' (1995) event duration scale is not as relevant in fire disturbance as it is in flooding. His rate scale is an important part of the fire disturbance story, but cannot be examined in the Heinselman example because the density of preserved information on fire extent is not constant through time.

A second approach to studying fire disturbance relies on remote sensing for either fire detection or determination of areal extent burned. In cases of fire detection the observational grain, the pixel, is commonly larger than the spatial extent of the event. Cahoon et al. (1999) point out that the nominal spatial resolution of the packed raster grid does not represent the area of energy recorded by a single pixel. They show that the point

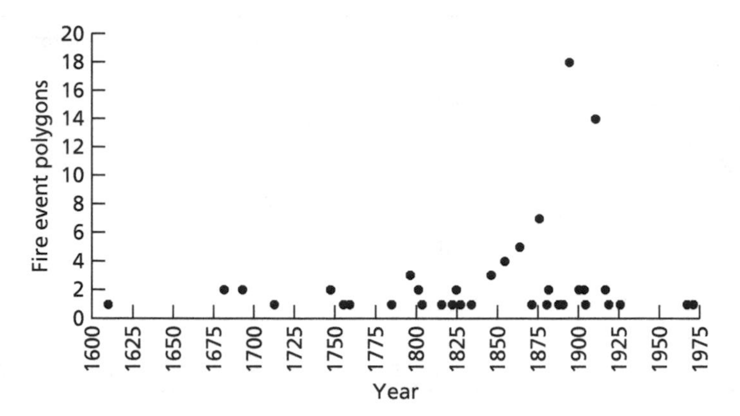

Figure 5.11 *Fire histories from fire scars recorded in tree rings for the Boundary Waters Canoe Area Wilderness of Minnesota.*
Source: Derived from maps by Heinselman (1973).

response function (PRF) is much larger for AVHRR images with a frequently reported nominal data portrayal of 1.1 km pixels. The full footprint is over 5 km on AVHRR for 99 percent of energy: 28 percent comes from nadir and 72 percent comes from outside the nadir pixel. That means that a hot fire in one nadir pixel stands a chance of being recorded by all adjacent pixels. Thus, there is a chance for data spreading. These traits must be considered when evaluating the fire model output. Midlatitude forest fires cover multiple pixels, while many savanna fires in places such as Angola are single pixel fires. Obviously, detection of these smaller fires is more problematic.

Fuller (2000) used three different detection thresholds in AVHRR channel 3 daytime images of Kalimantan, western Indonesia, to show the difference that threshold scales make. In such cases, the temperature resolution must be able to separate the fire averaged over a pixel from the ambient background. A 6 K higher threshold means an order of magnitude loss in fire class pixels. The spatial extent of Fuller's study was 13.18 million ha, represented at a spatial scale of about 1:10,000,000.

Dwyer et al. (2000) compiled remotely sensed global fire-detection maps for a single year divided into four seasons. The temporal extent is one year, with the potential for a daily observational time grain. The realistic repeat time for any one spot is greater than once per day due to clouds, and is geographically varied. The cartographic spatial scale of Dwyer et al.'s study is about 1:380 million at the equator, and their cartographic spatial grain is about 56 by 56 km. This coarse cartographic grain far exceeds the fire detection threshold. The cartographic time grain of three months is considerably coarser than the observational time grain.

A third approach to fire disturbance is fire modeling. It uses remote sensing only for verification. Scale plays an integral role in the modeler's selection of spatial and temporal grain size. The grain must be small enough to capture the spatial variability in flammability and its change through the time of the simulation. In tropical rainforests, the appropriate grain size depends on threshold gap size and the height of the canopy that govern the penetration of sunlight to the forest floor. Pereira (2002b) simulated fires to determine how large areas of flammability must be to enable a fire to span the entire map area. In these forests the fire process is strongly controlled by the size and spatial pattern of canopy gaps that result from logging. Pereira used a 30 m grain and an extent of 30 km so that the results can be compared with TM images. Such studies have implications for managing tropical forests only if the modeling captures the spatial and temporal grain and extent of the actual tropical forest fires.

Summary

Each biotic subject has an intrinsic spatial grain. Redwoods are larger than violets, and taxonomic families have a higher tolerance for within class dissimilarity than do species. Similarly, the geographic extent of a family is larger than the extent of a species in that family. Choosing the entities to observe is often a matter of convenience, time, or money.

Multiple biotic and abiotic environmental characteristics are analyzed together in the quest for convergence of evidence in hypothesis testing. Scale change is often performed in search of an explanation that does not exist in the grain or extent of the original data, or to match the spatial and grain scales of independent data sets. It is important to remember that the observational data with the greatest uncertainty establish the uncertainty of the entire analysis. Thus our confidence in biogeographical output should be calibrated by carefully evaluating the multiple embedded scales.

We hope that it is clear that the wide differences in embedded scales among maps in biogeography illustrate the importance of understanding how observations are made, and how they are transformed through geographic analysis to create maps or other graphics. Besides scales in time, space, and on maps, scales in attribute type and quantity affect the meaning of maps.

References

Allen, T. F. H. 1998: The landscape "level" is dead: Persuading the family to take it off the respirator. In D. L. Peterson and V. T. Parker (eds), *Ecological Scale*. New York: Columbia University Press, 35–54.

Benton, M. J. 1995: Diversification and extinction in the history of life. *Science*, 268:52–8.

Brown, D. A. and K. D. Brown 1996: Disturbance plays key role in distribution of plant species. *Restoration and Management Notes*, 14: 140–7.

Brown, D. A. 2002: *Alternative Biogeographies of the Global Garden*. 3rd edn. Department of Geography, University of Minnesota.

Cahoon, D. R. Jr, B. J. Stocks, M. E. Alexander, B .A. Baum, J. G. Goldammer 1999: Wild fire detection from space: theory and application. In J. L. Innes, M. Benston, and M. M. Verstraete (eds), *Biomass Burning and its Inter-Relationships with the Climate System.* Advances in Global Change Research 3, Dordrecht/Boston/London: Kluwer Academic Publishers, 151–69.

Ceballos, G. and P. Ehrlich 2002: Mammal population losses and the extinction crisis. *Science,* 296:904–7.

Clements, F. E., 1916: *Plant Succession: An Analysis of the Development of Vegetation,* Carnegie Inst. Wash. Pub No. 242:1–512.

Cook, E. R., S. W. Stahle and M. K. Cleveland 1992: Dendroclimatic evidence from eastern North America. In R. S. Bradley, and P. D. Jones, (eds), *Climate Since 1500*. London: Routledge, 331–48.

Corbet, G. B. and J. E. Hill 1992: *The Mammals of the Indomalayan Region: A Systematic Review*. New York: Oxford University Press.

Davis, M. B. 1983: Holocene vegetational history of the Eastern United States. In H. E. Wright, Jr., (ed.), *Late Quaternary Environments of the United States, vol 2 The Holocene,* Minneapolis, MN: University of Minnesota Press, 166–81.

De Cola, L. 1997: Multiresolution covariation among Landsat and AVHRR vegetation indicies, In D. A. Quattrochi and M. Goodchild (eds), *Scale in Remote Sensing and GIS*. Boca Raton, FL: CRC Press, 141–68.

Delcourt, H., P. Delcourt, and T. Webb, III 1982: Dynamic plant ecology: The spectrum of vegetational change in space and time. *Quaternary Science Review,* 1: 153–75.

Dwyer, E. S. Pinnock, J.-M. Gregoire, and J. M. C. Pereira 2000: Global spatial and temporal distribution of vegetation fire as determined from satellite observations. *International Journal of Remote Sensing,* 21: 1287–1302.

Franklin, J. and C. Woodcock 1997: Multiscale vegetation data for the mountains of Southern California: Spatial and categorical resolution. In D. A. Quattrochi and M. Goodchild (eds), *Scale in Remote Sensing and GIS*. Boca Raton, FL: CRC Press, 141–68.

Fredlund, G. 1994: Late Quaternary pollen record from Cheyenne Bottoms, Kansas. *Quaternary Research,* 43: 67–79.

Fuller, D. O. 2000: Satellite remote sensing of biomass burning with optical and thermal sensors. *Progress in Physical Geography*, 24: 543–61.

Gleason, H. A. 1926: The individualistic concept of the plant association. *Bulletin of the Torrey Botanical Club*, 53: 7–26.

Gravatt, Flippo 1949: Chestnut blight in Asia and North America. *Unasylva*, 3:2–7. Also available online at: http://www.fao.org/docrep/x5348e/x5348e02.htm. Accessed 11/01/01.

Heinselman, M. L. 1973: Fire in the virgin forests of the Boundary Waters Canoe Area, Minnesota. *Quaternary Research*, 3: 329–82.

Kingdon, J. 1997: *The Kingdon Field Guide to African Mammals*. New York: Academic Press.

Köppen W. 1931: *Grundriss der Klimakunda*, Berlin: Walter de Gruyer.

Küchler, A. W. 1949: Natural vegetation (map), In E. Espenshade, Jr., (ed.), *Goodes World Atlas*, Skokie, IL: Rand McNally.

Küchler, A. W. 1964: Potential natural vegetation of the conterminous United States. *American Geographical Society Special Publication*, 36: iii–116.

Loveland, T. R., J. W. Merchant, J. F. Brown, D. O. Ohlen, B. C. Reed, P. Olson, and J. Hutchinson 1995: Seasonal land-cover regions of the United States. *Annals of the Association of American Geographers*, 85: 339–55.

Marschner, Francis 1930: The original vegetation of Minnesota, (manuscript edited by Miron Heinselman and published by North-Central Forest Exp. Sta., U. S. Department of Agriculture, Forest Service in 1974).

Mitchell-Jones, A., G. Amori, W. Boganowicz, B. Krystufek, P. Reijnders, F Spitzenberger, M. Stubbe, J. Thissen, V. Vohralik, and J. Zima 1999. *The Atlas of European Mammals*. Academic Press, London.

Nicholson, Malcolm 1990: Henry Gleason and the individualistic hypothesis: The structure of a botanist's career. *The Botanical Review*, 56: 91–161.

O'Neill, R. V. 1998: Homage to St. Michael: Or, why are there are so many books on scale? In D. L. Peterson and V. T. Parker (eds), *Ecological Scale*, New York: Columbia University Press, 3–16.

Pan, Y. D., J. M. Melillo, A. D. McGuire, D. W. Kicklighter, L. F. Pitelka, K. Hibbard, L. L. Pierce, S. W. Running, D. S. Ojima, W. J. Parton, and D. S. Schimel 1998: Modeled responses of terrestrial ecosystems to elevated atmospheric CO_2: A comparison of simulations by the biogeochemistry models of the Vegetation/Ecosystem Modeling and Analysis Project (VEMAP). *Oecologia*, 114: 389–404.

Pereira, G. M. 2002a: A topology of temporal and spatial scale relations. *Geographic Analysis*, 34: 21–33.

Pereira, G. M. 2002b: Investigations of Fire Dynamics in Seasonally Dry Regions of the Amazon Rainforest. Unpublished PhD Dissertation, University of Minnesota.

Phillips, J. D. 1995: Biogeomorphology and landscape evolution: The problem of scale. *Geomorphology*, 13: 337–47.

Schimel, D., J. Melillo, H. Tian, A. D. McGuire, D. Kicklighter, T. Kittel, N. Rosenbloom, S. Running, P. Thornton, D. Ojima, W. Parton, R. Kelly, M. Sykes, R. Neilson, and B. Rizzo 2000: Contribution of increasing CO_2 and climate to carbon storage by ecosystems of the United States. *Science,* 287: 2004–6.

Schoomaker, P. K. 1998: Paleoecological perspectives on ecological scale. In D. L. Peterson and V. T. Parker (eds), *Ecological Scale,* New York: Columbia University Press, 79–103.

Shantz, H. L. and R. Zon 1924: Natural vegetation (of the United States), *Atlas of American Agriculture,* Washington, D.C.: U.S. Department of Agriculture.

Thomas, I. 2001: *Untitled,* caribou migration map. This map was an ephemeral web posting of caribou migration map, USGS/NBS.

Turner, M. G., R. H. Gardner, R. V. O'Neill 2001: *Landscape Ecology in Theory and Practice,* New York: Springer-Verlag.

VEMAP Members 1995: Vegetation/Ecosystem Modeling and Analysis Project (VEMAP): Comparing biogeography and biogeochemistry models in a continental-scale study of terrestrial ecosystem responses to climate change and CO_2 doubling. *Global Biogeochemical Cycles,* 9(4): 407–37.

Ziegler, S. S. 2000: A comparison of structural characteristics between old-growth and postfire second-growth hemlock-hardwood forests in Adirondack Park, New York, U.S.A. *Global Ecology and Biogeography,* 9: 373–89.

6 Scaled Geographies: Nature, Place, and the Politics of Scale

Erik Swyngedouw

Nature, Place, and Scale: A Historical-Materialist Perspective

In early 1998 (*Le Monde*, 17 January), controversy arose in the Paris region about IBM's continuing tapping of ancient underground aquifers. The production of new generation computer chips requires large volumes of water of the highest purity to cleanse micropores. Environmentalists, seeking to protect historical "natural waters," were outraged. The water company, Lyonnaise des Eaux, was worried about the potential loss of water and, consequently, future dividends. The state at a variety of scales was caught up in the myriad of tensions ensuing from this: protection of the natural environment versus economic priorities, the competing claims of different companies, etc... The ancient underground waters fused with politics, economics, and culture in intricate ways.

This is just one example from a proliferating number where the traditional distinction between environment and society, between nature and culture, becomes blurred, ambiguous, and problematic. The contested "making" of "Dolly," the cloned sheep, the outbreak of BSE (mad cow disease), the built-up of CO_2 in the atmosphere, and the depletion of ozone in the stratosphere similarly fuse physical–environmental metabolisms with sociocultural and political–economic relations. These all suggest how nature and society are constituted as networks of interwoven processes that are human and natural, real and fictional, mechanical and organic. They also suggest how the social and physical transformation of the world is inserted in a series of scalar spatialities. "Dolly," Ozone, or Parisian aquifer waters all embody and express physical and social processes, whose drivers operate at a variety of interlocked and nested geographical scales.

This chapter addresses the scalar construction of socionatural processes and the centrality of a politics of scale in the production of particular geographical configurations. This problematic will be approached from a historical-geographical materialist perspective. First, I examine the question of nature, place, and scale. Second, two examples of the contested construction of spatial scales, in which the social and natural operate in inseparably intertwined manners, are presented. Finally, the importance of a radical politics of scale in the construction of emancipatory political agendas and strategies is discussed.

On nature

In recent years, there has been a resurgence of historical-materialist thought on nature (Benton, 1996; Castree, 1995; Grundman, 1991; Harvey, 1996; Hughes, 2000; Smith, 1984; Swyngedouw, 1999a). Historical-geographical materialism is founded on the ontological principle that living organisms, including humans, need to transform (metabolize) "nature" and, through that, both humans and "nature" are changed. Marx hastens to add that this metabolic transformation of nature (environmental change) is always a social and historical process. Put simply, in order to live, humans transform the world they live in, and this takes place in interaction with others; that is under specific "social relations of production." This metabolism is necessarily a social process. Both nature and humans, materially and culturally, are profoundly social and historical from the very beginning (Smith, 1996; 1998; Castree, 1995; Haraway, 1997). Although early Marxists tended to focus on questions of distribution and power among and between humans and social groups, the inevitable physical transformation of nature and the production of new "natures" (both materially and socially) remained as a presupposition. The social appropriation and transformation of nature produces historically specific social and physical natures that are infused by a myriad of social power relationships (Swyngedouw, 1996a). Social beings necessarily produce nature; nature becomes a sociophysical process infused with political power and cultural meaning (Haraway 1991; 1997). In addition, the transformation of nature is embedded in a series of social, political, cultural, and economic constellations and procedures (i.e., social relations) that operate within a nested articulation of significant, but intrinsically unstable, geographical scales.

On place and space

The process of perpetual metabolic transformation of social and physical nature and the transformation of social life are part and parcel of the same process. Everyday life is necessarily "placed" or "situated" by virtue of the need to transform and metabolize (produced) nature. The material and social conditioning of life and of the metabolic transformation of nature is constituted in and through temporal/spatial social relations that operate over a certain scalar extent. Engaging place as "produced" nature is essential for human existence (Swyngedouw, 1997a). Under capitalism, place as (produced) nature (socially transformed or given) becomes a central element in the forces of production that shape and partly condition accumulation trajectories and strategies (Swyngedouw, 1992). At the same time, place embodies a historical layering of crystallized social relations.

The process of the production of place/nature is inevitably a contradictory one as it necessarily implies a process of "creative destruction" of nature/society. The conflicting (capitalist) social power relations (along class, gender, or other social cleavages) through whom this transformation is organized perpetually destroy existing conditions to replace them with new configurations and characteristics. "Creative destruction" is always an already social process: the process of metabolic transformation of produced nature takes place in association with others. The thing that is transformed and the thing that arises out of the transformation process is always already part of and embodies the social relations through which nature/society is transformed. The world's historical geography can, consequently, be reconstructed from the vantage point of this perpetual socioecological transformation process. These social relations are always constituted through temporal and spatial relations of power with respect to the social and physical ecology that is being transformed. Indeed, these social relations are "grounded" in the sense that they regulate (but in highly contested or contestable ways) control over and access to transformed nature (place), but these relations also extend over a certain material/social space. They produce what Massey (1993) refers to as a "geometry of power." It is also here that the issue of geographical scale emerges as central. Sociospatial relations operate over a certain distance and produce scalar configurations.

Scaled geographies: scaling nature – scaling the social

I insist that social life is process-based, in a state of perpetual change, transformation and reconfiguration (see Harvey, 1996). Starting analysis from a given geographical scale seems to me, therefore, to be deeply antagonistic to apprehending the world in a dynamic, process-based manner. This has profound implications for what scale means. I conceive scalar configurations as the outcome of sociospatial processes that regulate and organize social power relations. As a geographical construction, scales become arenas around which sociospatial power choreographies are enacted and performed (Swyngedouw, 1997a; 1997b; 2000b). Over the past few years, a plethora of research has been published on the social construction of scale and the deeply contested scalar transformations of the political-economy of advanced capitalist societies (Dicken et al., 2001; Howitt, 1993; Smith and Dennis, 1987; Swyngedouw 1997a; 1997b; 1998). Emphasis has been put on the making and remaking of social, political, and economic scales of organization (Brenner, 1998; Collinge, 1999; Cox, 1998; Delaney and Leitner, 1997; MacLeod and Goodwin, 1999; Marston, 2000: Marston, this volume; Silvern 1999), of regulation (Boyle, 2000; Berndt, 2000; Brenner, 1997; Leitner, 1997; Swyngedouw, 1992), of social and union action (Herod, 1991; Sadler, 2000; Walsh 2000), and of contestation (Castree, 1999; Miller, 1997; Towers, 2000). In addition, attention has been paid to the significance of differential scalar positionings of social groups and classes in the power geometries of capitalism (Kelly, 1999; MacLeod, 1999; Swyngedouw, 2000a; Leitner, this volume), and on scalar strategies (the jumping of scales) mobilized by both elites and subaltern social groups (Brenner, 1999; Herod, 1991; Swyngedouw 1996b; Zeller, 2000; Smith, this volume).

With a few notable exceptions, the question of nature has remained largely outside this analysis (Escobar, 2001; Grainger, 1999; Zimmerer, 2000). I insist that nature and environmental transformation are also integral parts of the social and material production of scale. More importantly, scalar reconfigurations also produce new sociophysical ecological scales that shape in important ways who will have access to what kind of nature, and the particular trajectories of environmental change. The examples in the next section attempt to substantiate and elucidate how the "scaling of nature" is deeply intertwined with the scaling of social life and of the power relations inscribed therein. Before we embark on this, I recapitulate my perspective on the social and material production of scale and scalar gestalts:

1 Scalar configurations, whether ecological or in terms of regulatory order(s), as well as their discursive and theoretical representation, are always already a result, an outcome of the perpetual movement of the flux of sociospatial and environmental dynamics. The theoretical and political priority, therefore, resides never in a particular geographical

scale, but rather in the process through which particular scales become constituted and subsequently transformed.

2 Struggling to command a particular scale in a given sociospatial conjuncture can be of eminent importance. Spatial scales are never fixed, but are perpetually redefined, contested and restructured in terms of their extent, content, relative importance and interrelations. The continuous reshuffling and reorganization of spatial scales are integral to social strategies and an arena for struggles for control and empowerment.
3 A process-based approach to scale focuses attention on the mechanisms of scale transformation through social conflict and political-economic struggle. In many instances, this struggle pivots around the appropriation of nature and control over its metabolism. These sociospatial processes change the importance and role of certain geographical scales, reassert the importance of others, and on occasion create entirely new scales. These scale redefinitions in turn alter the geometry of social power by strengthening the power and the control of some while disempowering others (see also Swyngedouw, 1989; 1997b; 2000a).
4 Smith (1984) refers to this process as the "jumping of scales," a process that signals how politics are spatialized. That is, scalar political strategies are actively mobilized as parts of strategies of empowerment and disempowerment. As the scalar "gestalt" changes, the social power geometry within and between scales changes.
5 There is a simultaneous, "nested" (like a Russian doll), yet partially hierarchical, relationship between scales (Jonas, 1994: 261; Smith, 1984; 1993). Clearly, social power along gender, class, ethnic or ecological lines refers to the scale capabilities of individuals and social groups. Engels (1844) already suggested how the power of the labour movement, for example, depends on the scale at which it operates, and labour organizers have always combined strategies of controlling place(s) with building territorial alliances that extend over a certain space.
6 Scale configurations change as power shifts, both in terms of their nesting and interrelations and in terms of their spatial extent. In the process, new significant social and ecological scales become constructed, while others disappear or become transformed.
7 Similarly, ecological scales are transformed as and when the socioecological transformation of nature takes new or different forms. For example, the multiscalar configurations of monocultural cash-cropping agriculture are radically different the socioecological scales of peasant subsistence farming.
8 Scale also emerges as the site where cooperation and competition find a (fragile) standoff. For example, national unions are formed through alliances and cooperation from lower scale movements, and a fine

balance needs to be perpetually maintained between the promise of power yielded from national organization and the competitive struggle that derives from local loyalties and interlocal struggle.

9 Processes of scale formation are cut through by all manner of fragmenting, divisive and differentiating processes (nationalism, localism, class differentiation, competition, and so forth). Scale mediates between cooperation and competition, between homogenization and differentiation, between empowerment and disempowerment (Smith, 1984; 1993).

10 The mobilization of scalar narratives, scalar politics, and scalar practices, then, becomes an integral part of political power struggles and strategies. This propels considerations of scale to the forefront of both ecological and emancipatory politics.

In sum, the condition of everyday life resides in the twin condition of the essential transformation of nature (place) on the one hand and sociospatial relations through which this transformation is organized and controlled on the other (Swyngedouw, 1992). It is exactly this process that Lefebvre (1991) refers to as "The Production of Space" and it involves the production of scalar or scaled geographical configurations. The geometries of power, of course, fragment and differentiate them in multiple ways as I attempt to illustrate below.

The World in a Cup of Water: Scalar Processes and the Contested Politics of the Rescaling of H_2O

I briefly examine two cases, which use water as the conceptual and material entry into a particular aspect of the social and material production of scale, the making of scalar articulations, and the politics of rescaling. Life is hardly imaginable without water. The multiple temporalities and interpenetrating circulations of water (the hydrological cycle, canalization, and distribution networks of all kinds, dams, etc.) illustrate its perpetual physical and social metabolism and mobilization. Water relates all things/subjects in a network, or rhizome, connecting the most intimate of sociospatial relations; and inserts them in a complex political-economy and political–ecology of bodily, local, urban, regional, national, and international scales. Circulating water also is part of a chain of local, regional, national and global social and ecological flows of H_2O, money, texts, and bodies.

We can use water as an entry-point to reconstruct, and hence theorize scalar transformations as a political-ecological process. Water embodies, simultaneously and inseparably, biochemical and physical properties, socioeconomic and political characteristics, and cultural and symbolic meanings. These multiple metabolisms of water are structured and organized through relations of power, that is relations of domination and subordination, of access and exclusion, of emancipation and repression. This circulation of water is embedded in and interiorizes a series of multiple power relations along ethnic, gender, and class lines. These situated power relations, in turn, swirl out and operate at a variety of interrelated geographical scale levels, from the scale of the body upward to the political-ecology of the city and its hinterland, and to the global scale of uneven development. The struggle over nature and the uneven access to water turns the issue into a highly contested terrain that captures wider processes of political–ecological change.

My first example demonstrates how urbanization itself involves the continuous reconstruction of social and ecological scales, while producing new scalar configurations. This will be developed through a brief historical geography of the urbanization of water in Guayaquil, Ecuador (Swyngedouw 1995; 1997c). The second example illustrates how the mobilization of a particular scientific discourse on a specific physical scale (the river basin) becomes an arena for staging political power choreographies that were decisive in shaping processes of modernization in Spain (Swyngedouw, 1999b). This shows how "scales of nature" become incorporated into particular political projects.

Conflict, scale, and the urbanization of H_2O in Guayaquil, Ecuador

Guayaquil, Ecuador's largest and most powerful city located on the Pacific coast, suffers from a seriously socially uneven access to potable urban water, like many other cities in developing countries. Of its two million inhabitants, 38 percent do not have access to piped potable water, and depend on private vendors who sell water at a massively inflated price. Publicly supplied water costs approximately three cents for 1000 litres, while private water vendors charge three dollars. As a result, an intense social and political struggle, enacted at bodily, neighbourhood, urban, regional, national, and international scales, unfolds over access to and control over the city's water resources. The uneven power relationships that have shaped Guayaquil's urbanization process are thus etched into the circulation of urban H_2O. The historical geography of the urbanization of water suggests how particular physical-ecological, political, and economic scales are constructed and perpetually reconstructed. It also shows how the resulting scalar configurations become nested arenas for further social and political struggle over access to water.

Clearly, the urbanization process itself is predicated upon the mastering and engineering of nature's water. The ecological conquest of water is, therefore, an integral part of the expansion and growth of the city. At the same time, the capital required to build and expand the urban water landscape itself is, at least in the case of Guayaquil, generated through the political-ecological transformation of the city's hinterland and the successive incorporation of both expanding water volumes as well as new forms of socioecological metabolism. The city's growth has required a progressive geographical expansion of its water footprint. As more migrants flocked to the city, water systems had to move further away from the city in search of new or additional water resources. Simultaneously, the financing of these capital-intensive projects, whose technology was invariably imported from abroad, necessitated the generation of sufficient foreign currency and, consequently, a sound export-based economy. These capital flows were generated initially on the basis of cocoa (circa 1890–1930), followed by bananas (circa 1950–1970) and oil (after 1972). With each successive phase, the scalar configurations of power at the local, regional, national, and international level became transformed and rearticulated. In what follows, we shall explore the historical dynamics of the urbanization process through the lens of this double ecological conquest.

At the turn of the twentieth century, the city's elites mobilized around a growing preoccupation with the presence and role of water in the city. This paralleled a changing sociospatial class position and a reconfiguration of the state apparatus. After independence (1830) and particularly after 1850, the early postcolonial society underwent significant sociospatial changes as Ecuador was gradually transformed into an agro-export economy. This Ecuadorian accumulation model originated with the expansion of world demand for and trade in cocoa around 1860. Cocoa accounted for 90 percent of total exports by 1890, and in 1904 Ecuador became the world's leading cocoa exporter (Aguirre, 1984; Chiriboga, 1980: 261). The coastal socioecological complex, originally mainly characterized by small-scaled and a largely self-contained peasantry, had given way to immense cocoa plantations involving a variety of forms of waged work. The forced and rapid formation of a wage-dependent class, combined with a fast depeasantization process, fed growing demands not only for wage labour in the coastal plantations, but also for auxiliary waged functions in the city. Between 1896 and 1920, Guayaquil grew from 50,000 to 100,000 inhabitants (Rojas and Villavicencio, 1988: 22).

The rise of the emergent Guayaquileño metropolis was predicated on the transformation of nature and the integration of a new cocoa-based agricultural ecology in the process of commodity production and rent extraction. Countryside and city were both restructured through this socioecological conquest, which inserted the central coastal region of Ecuador squarely into a worldwide money-circulation process and produced the city as the nexus for rent appropriation and distribution. At the same time, the spatial scaling of political power was also redrawn. Through these political-economic and ecological shifts, the urban merchant bourgeoisie, in alliance with coastal landowners and cocoa producers, now controlled the city and the countryside and began to aspire for more national political influence. The coastal political elite increasingly challenged the hegemony of the traditional highland (Serrano) landed "aristocracy" (Guerrero, 1980). Eventually, the coastal "cocoa" elite managed to "jump scale" and displace the highland aristocracy from the helm of the national state apparatus.

In 1900, Eloy Alfaro, Guyaquileño politician and president of Ecuador, declared the urban water project and other sanitary infrastructure a work of national importance, to be financed largely by the national state on the basis of taxes levied on cocoa exports. Between that moment and the 1930s, the urban water system was gradually extended, following, but lagging behind, the pace of urbanization. It became evident that the water frontier needed to be pushed outward in search of new exploitable water reserves, in order to redress the imbalance. The growth and expansion of the city could only be sustained by incorporating ever-larger parts of nature's geography into the circulation of money and profit upon which the city's continuing prominence crucially depended. This incorporation of new "natural" waters into the urban water circulation process then enabled the extension of the material scale of the urban network.

This successful watering of the city was very short lived however. The urbanization of water slowed down dramatically as political power relationships began to shift in decisive new ways, particularly after the crumbling of the cocoa economy. By the end of the 1930s, the highly successful and hegemonic bourgeois growth coalition that had launched Guayaquil on a path of dependent modernization had fallen apart. The collapse of the cocoa economy produced the first cracks in the hitherto firmly allied coastal-regional elite alliance of cocoa producers, merchants, and financiers. The socioecological opening up of Africa for world cocoa production, phytosanitary problems resulting from monocultural practices, and a dwindling demand for cocoa from Europe during the World War I negatively affected prices, productivity, and production. Cocoa revenues fell by 21 percent between 1917 and 1926, and cocoa output fell by 45 percent, from 1,008,000 to 447,000 quintals (Bock, 1988: 60). The urbanization of water stuttered during this period. Changing socioecological processes in the urban region were thus, in a myriad of intricate ways, related to and expressive of fluctuations on the New York commodities exchange market and the vagaries of the international monetary system.

The disintegration of the cocoa economy threw many agricultural semi-proletarian workers into unemployment and poverty, fuelling a mass migration to the city. The city experienced rapid population growth (182 percent between 1925 and 1950), mainly through urban land-invasions and the construction of informal settlements by impoverished former cocoa workers.

While the urban population expanded, the urbanization of capital dried up, including investments in collective infrastructure. The resulting slowdown in the urbanization of water in the context of an expanding population led to an acute water crisis by the end of the forties. Water problems would never really go away again. On the contrary, exclusionary water politics and water speculation by vendors would increasingly characterize urban struggles, becoming integral to the rituals of everyday urban life.

The turbulent but lean years of the 1940s were followed, however, by the banana bonanza decade of the 1950s. The United States' fruit corporations, their plantations devastated by Panama disease, moved their centre of operations from Central American and Caribbean exporters to Ecuador. It was a cheap location, and the Panama disease had not moved that far South. The subsequent spiralling demand for bananas converted the coastal area of the country (La Costa) into large banana plantations with their associated socioecological relations (Armstrong and McGee, 1985: 114; Larrea-Maldonado, 1982: 28–34; see also Schodt, 1987). Banana export receipts exploded from US$ 2.8 million in 1948 to $ 88.9 million in 1960, accounting for 62.2 percent of Ecuador's total exports (Hurtado, 1981: 190; Grijalva, 1990; Cortez, 1992). This manufactured "banana-bonanza" was organized through a new political-economic and ecological transformation. The ecological frontier for agricultural export production around Guayaquil was pushed further inland (León, 1992; Trujillo, 1992), radically altering the scalar social and physical ecology of the urban–rural complex and incorporating ever-larger areas into the global circulation of money. Although smallholdings predominantly organized actual production, its commercialization was concentrated in very few hands, combining a tiny regional-national comprador elite with US global fruit-trading companies (Báez, 1985). This banana colonization prompted mass migration to the coastal areas, catalysing further rapid growth of Guayaquil, whose banana-dependent financial and service economy expanded rapidly (Carrión, 1992). Between 1950 and 1974, the city's population grew from 200,000 to over 820,000.

Banana rents were ploughed back into the urban realm, either directly or indirectly through the state (Báez, 1992). The backbone of Guayaquil's accelerated urbanization process was rooted in the expanded and reworked ecological conquest of the coastal region, and nested in an expanding metropolitan and global agro-business complex. Economic growth improved Ecuador's credibility and, helped by the efforts of the newly established international financing organizations, foreign capital again began to flow into Ecuador. This fine-grained texture of economic, political, social, and ecological transformations produced a ferment from which the postwar expansion of the urban water frontier to new and hitherto unexploited water reserves would emerge. Banana rents were combined with international loans to finance rapid urbanization (and peripheral modernization) of the country. This new ecological conquest combined with a reinvigorated quest for control over and domestication of nature's water.

In 1947, a new source for drinking water for Guayaquil surfaced as the next target to harness, the river Daule, but it would take until the 1950s banana boom before these plans could be realized. Together with its expanding role as a water source for irrigation projects in the region, the flow of the Daule was to be diverted, transformed, and commodified. Banana export earnings, combined with a reverse flow of money from the US, were welded together with the flow of Daule water to circulate through the veins of the city, reshaping its landscape. But this material flow of H_2O, combined with and running through physical and social urban space, was just one node in an articulated whole of processes operating on a regional, national, and, indeed, world-wide scale: flows of transformed nature, commodities (bananas), and money; transfers of capital; and the buying and selling of labour power (see Merrifield, 1993). The city would be transformed once more, with the political-economy of urbanization deeply caught up in the progress of the urbanization of water.

This new scalar configuration of the water/banana nexus came to an early end beginning in the early 1960s. In the 1950s, a new and more resistant banana variety, the Cavendish, was developed, allowing the fruit companies to switch their operations back to the more favourably located Central American locations, closer to "home," more reliable and under greater direct control of the US state. This bioengineered and phyto-technologically more demanding "Chiquita" banana (León, 1992) was heavily commercialized internationally and undermined the economic position of the traditional Ecuadorian "Gross Mitchel" banana type. Only large Ecuadorian producers, connected to international merchants and fruit companies, were able to adjust ecologically and socioeconomically to the requirements of the new cultivation, production, and marketing techniques. Output continued to grow until by the early 1960s production was twice the exported volume. International merchants could be more selective and demanding. Total banana export value fell from US$ 88.9 million in 1960 to $ 51.5 million in 1965, recovering (nominally) to $ 94.3 million by 1970 (Tobar, 1992: 238). This overaccumulation of bananas wiped out thousands of small and medium sized producers, who joined the ranks of the urban underclass (Bàez, 1985: 554). The banana crisis again broke the coastal elite's partially restored power position. The state, in turn, was pushed to face the stagnant export position of Ecuador, as external debt rose rapidly.

The exploitation of Amazonia's huge oil reserves in eastern Ecuador after 1972 signalled a new wave of rent extraction and redistribution. Existing sociospatial and scalar relations were overhauled once more, as the actors organising the petroleum boom produced a new set of scalar configurations. The ecological conquest of fossilized nature beneath the Ecuadorian Amazonian rainforest was, and is, exclusively based on international petro-capital. In contrast to the two earlier waves of agro export-based integration into the international market place (cocoa and bananas), mainly organized through the intermediation of a domestic commercial and financial oligarchy, this time the national state assumed the role of key interlocutor in organising the global-local articulation of oil. Indigenous Amazonian peoples were legally dispossessed, as the state became the *de facto* and *de jure* owner of the country's "natural" resources (Báez, cited in Farrell, 1989: 146). This would, of course, put the state in the pole position in terms of organising the insertion of Ecuador into the global political economic framework, inevitably also turning the state apparatus into a major arena for social struggle. Oil revenues, partly monopolized by the state, triggered continuous political power conflicts over the control, appropriation, and direction of the new investments that now became possible. In addition, the oil boom attracted considerable attention from foreign investors (mainly in services and banking). The majority of this private investment was increasingly attracted to the inland capital city of Quito, rather than Guayaquil, which had the advantage of proximity to key national and international power brokers.

This time, the expansion of the ecological rent frontier was directed eastward into the Amazon basin rather than in the coastal regions. Oil, quite literally, flowed to the coastal port (for export) over the Andes through a newly constructed oleoduct, becoming transformed into money and capital. Quito became the country's leading political and now increasingly international financial centre, leaving Guayaquil behind in its past, but now dimmed, glory. The oil rents appropriated by the state were reinvested, in turn, with an eye toward domestic industrialization (Bocco, 1987), mainly in all sorts of infrastructure, from expanding port facilities, new freeways to airports, and a military built-up. Oil rents also served to augment the ecological basis on which the city's sustainability was predicated, including widening the scale and scope of water control. The pumping, treatment, and conduction capacity of Guayaquil's water system was increased substantially (reaching 1,500 million m^3 in 1995), taking ever more water from the Daule river and its tributaries. The expansion of the water system was largely financed from international loans, secured by promises of a continuing oil boom, but a significant part of the urban population was deprived from easy access to potable water. The socio-economic crisis of the 1980s had led to a massive explosion of the city to over two million people, particularly in marginal estuary settlements and on the hills surrounding the old city. The lack of attention to water distribution and the absence of a piped network resulted in chronic problems of access to water for the urban poor and fostered a thriving private water economy.

To summarize, the city of Guayaquil grew on the basis of successive ecological conquests and the appropriation of rents, from agricultural produce or the pumping of oil, through which money was continuously recycled and nature became urbanized. The harnessing and urbanization of water inserted water circulation squarely into the circulation process of money and its associated power relations and class differentiations. With each round of accumulation, the territorial scale of the socioecological complex changed and the scalar geographies of political power became rearticulated. The socioeconomic, political, and institutional scalar nesting (from the local to the global) through which cocoa, bananas, and oil (either in a commodity or money form) flowed took new forms. In addition, the scalar choreography of water circulation became transformed and restructured, and expressed and reflected the changing social, political, and economic power relations at a variety of nested and articulated geographical scales; urban, regional, national, and international.

Modernity, fascism, capitalism and the contested scaling of H_2O in twentieth-century Spain

Spain's history of modernization has been one of altering, redefining, and transforming the physical characteristics of its landscape and, in particular, its waterscape. Today, the country has almost 900 dams, more than 800 of which were constructed during the second half of the twentieth century. Every single river basin has been altered, managed, engineered, and transformed. Water has been an obsessive theme in Spain's national life during the last century and the quest for water continues unabated (del Moral Ituarte, 1996; 1998). Understanding the construction of a particular set of nested scales, and the mobilization of specific spatial scales by particular social groups, is necessary to grasp the choreographies of power and the strategies deployed to push through this modernising project. This process was rife with intense conflict: socioeconomic and political disintegration during the first decades of the twentieth century, a bloody civil war placing modernization under the control of a fascist dictatorship until 1974, and subsequent rapid transformation into a liberal democracy. In this example, we shall show how the conflict between modernizers and traditionalists took the form, among others, of a struggle over making and controlling the scale of river basin authorities.

Beginning in the late nineteenth century, the modernising desires of an emerging intellectual elite of "regeneracionists" crystallized around the transformation of Spain's hydrological structure, in an attempt to harness Spain's waters as the foundation for its economic and political revival (see Swyngedouw, 1999b). Water rapidly became a prime consideration in national political, socioeconomic, and cultural debates. Spain found itself in a traumatic condition at the turn of the twentieth century, having lost its last colonial possessions (Cuba and the Philippines) exactly when other imperial countries were consolidating their empires, and its internal political, economic, and social conditions were rapidly deteriorating. Unable to found Spain's modernization on an external geographical project of scale-enlargement, Spanish modernising elites concentrated on an equally geographical national program, but founded on the radical transformation of Spain's internal geography – particularly its water resources (Gómez Mendoza and Ortega Cantero, 1987). As Joaquin Costa, a regeneracionist intellectual, argued in 1880: "[I]f in other countries it is sufficient for man to help Nature, here it is necessary to do more; it is necessary *to create her*" (Costa, cited in Driever 1998: 40, author's emphasis).

This concern was also voiced by others (like Lucas Mallada (1890) or R. Macías Picavea (1899)). This program of producing a new sociophysical space embodied physical, social, cultural, moral, and aesthetic elements, fusing them around the dominant and almost hegemonic ideology of national development, revival, and progress.

The hydraulic intervention to create a waterscape supportive of the modernising desires of the regeneracionists, and of the social and political foundations of the existing class structure and social order, was very much based on a respect for "natural" laws and conditions. The latter were thought to be intrinsically stable, balanced, equitable, and harmonious. The hydraulic engineering mission thus consisted primarily in "restoring" the "perturbed" equilibrium of the erratic hydrological cycles in Spain. Of course, this endeavour required significant scientific and engineering enterprise, in terms of understanding and analysing nature's "laws," and in using these insights to work toward a restoration of the "innate" harmonious development of nature. The moral, economic, and cultural "disorder" and "imbalances" of the country at that time were seen as paralleling the "disorder" in Spain's erratic hydraulic geography, and both needed to be restored and rebalanced.

Two threads need to be woven together in this context: the pivotal position of a particular group of scientists, the Corps of Engineers (Villaneuva Larraya, 1991), and changing visions about the scientific management of the terrestrial part of the hydrological cycle. Both were linked to the rising prominence of hydraulic issues on the sociopolitical agenda at the turn of the century. The Corps of Engineers, founded in 1799, remains the professional collective responsible for the development and implementation of public works. It is a highly elitist, intellectualist, "high-cultured," male-dominated, and socially homogeneous and exclusive organization that has taken a leading role in Spanish politics and development (Mateu Bellés, 1995).

In line with the then emerging scientific discourse on orography and river basin structure and dynamics, the engineering community argued for a technical, political, and managerial intervention on the basis of the "natural" integrated water flow of watershed regions, rather than on the basis of historically and socially formed administrative regions (see Figure 6.1). This plea for an orographic regionalization overlaid the traditional political-administrative divisions of the country, forcing a reordering of the territory on the basis of its river basin structure. The engineers portrayed the latter as the crucial planning unit and political scale for hydraulic interventions. Cano García (1992) succinctly summarizes this scientific perspective:

> To revert to the great orographical delimitation for organising the division of the land represents a contribution made from within the strict field of our discipline [engineering] and at the same time, at least initially, it shows the abandoning of traditional political divisions and the importance of other perspectives and concepts. (Cano García, 1992: 312, author's translation)

As T. Smith (1969: 20) argues, . . . "the identity of the drainage basin seemed to offer a concrete and "natural" unit which could profitably replace political units as the areal context for geographical study." Brunhes (1920: 93) insisted on the water basin as the foundation for the organization of the land since "water is the sovereign wealth of the state and its people" (see also Chorley, 1969). Such a view was widely recounted in Spain at the time, and its arguments were rallied in defence of a new orographic-administrative organization of the territory.

This "scientific" and "natural" division, based on the spatial scale of the river basin, provided an apparently enduring and universal scale for territorial organization in lieu of the historically more recent and "constructed"

Figure 6.1 *Watershed regions in Spain.*

political scales associated with politico-administrative boundaries. The history of the delimitation of Hydrological Divisions based on the river basin is infused with the influence of the modernising hydraulic discourse, on the one hand, and the "scientific" insights gained from hydrology and orography on the other. The attempt to "naturalize" political territorial organization was part and parcel of a strategy of the modernizers to challenge existing social and political power geometries. The construction of and command over a new territorial scale might permit them to implement their vision and by-pass more traditional and reactionary power configurations. Indeed, the older and historically constructed administrative political scales (municipality, province, and nation-state) were firmly under the hegemonic control of traditional semifeudal elites who held a tight grip over society and resisted the structural transformations called for by modernizers.

Capturing the scale of the river basin as the geographical basis for exercising control and power over the organization, planning, and reconstruction of the hydraulic sphere was one of the central arenas through which the power of traditionalists (and the scales over which they exercized control) was challenged. River basins became the scale par excellence through which the modernizers tried to erode the powers of the more traditional provincial or national state bodies, while traditional elites held to the existing administrative territorial structure of power. The bumpy history of the hydrological divisions records this struggle (Gómez Mendoza and Ortega Cantero, 1992).

This negotiation of scale and the science/politics debate around the scaling of hydraulic intervention and planning raged for almost a century, before the current structure of river basin institutions was put into place (Cano Carcía, 1992; Mateu Bellés, 1995). The Water Act of 1879 had established that all surface water was common property, managed by the state. This also implied the need to create administrative structures to perform these managerial tasks (Giansante, 1999). The first Hydrological Divisions (ten in total) were established by Royal Decree in 1865, and were considered from the beginning to be major instruments for economic modernization. Some of these divisions more-or-less coincided with major river basins (Ebro, Tajo, Duero), others (particularly in the South) had a much closer correspondence to provincial boundaries. All were named after the provincial capital city where the head-office was located (Mateu Bellés, 1994). Their basic merit in those early days was to serve as an institutional basis for collecting statistical data to assist research into the hydrological cycle. These surveys could then be used as inputs to the real power holders: provincial head offices for public works, special ad hoc commissions, or private industry (del Moral Ituarte, 1995). The ten hydrological divisions were abolished in 1870, partly re-erected a few months later, reduced to seven, abolished again in 1899, and re-established in 1900 when their tasks extended to include the detailed study and planning of, and the formulation of proposals for hydraulic interventions. However, the ultimate decision-making power would remain with the traditional provincial level, which supervised and executed hydraulic works, and with the central state, for financing and controlling the infrastructure programs (Mateus Bellés, 1994). Control by the conservative local and national state fatally stalled implementation of these projects.

The complex and perpetually changing administrative organization and power structures associated with the successive attempts to establish river basin authorities, and their relative lack of power until the 1930s, reflect the failure of the early modernizers to successfully challenge traditional power lineages and scales (Mateu Bellés, 1994; Mateu Bellés, 1995). Only after 1926 were the current Confederaciones Sindicales Hidrográficas gradually established as quasi-autonomous organizations in charge of managing water, as stipulated by the 1879 Water Act (Giansante, 1999). The last of these ten Confederaciones was finally established only in 1961 (see Figure 6.1). What had proven impossible to achieve during the first decades of the century was finally fully implemented during the Franco dictatorship. Franco's fascist rule permitted the final formation of powerful river basin authorities, and aligned the national state more closer with the interests of the engineering community in reorganising the hydraulic geography of the country. The Confederaciones acquired a certain political status with participation from the state, banks, Chambers of Commerce, provincial authorities, etc. At each stage engineers took leading roles and became the activists of the regeneracionist project through a combination of their legitimization as holders of scientific knowledge and insights, and their privileged position as a political elite corps within the state apparatus.

By the end of Franco's rule, in 1974, Spain's hydroscape had been overhauled profoundly. Every single river basin is now fully managed to the "last drop" of available water. With the advent of democracy, however, the politics of scale around the water nexus took a new twist, as the ongoing desire to modernize the Spanish economy required ever-greater control over and management of the country's available water resources. As limits to river basin-based water management became evident, the water engineering community and its socioeconomic allies "jumped scales" and began to argue and lobby for the material construction of a national water-grid. The latter would produce a national water system, connecting every river-basin to form a national managerial and material (infra)structure. This would permit significant interbasin water transfers and a more "efficient" use of the available water resources. Over the past twenty years, this national water project has become a major domain of political conflict, in what is now a liberal-democratic polity. Various spatial scales, such as regional interests, localist strategies, and national projects, have faced-off against each other. Different social groups, such as ecologists, the agricultural lobby, the tourist industry, the energy sector, and regionalists also mobilize different scales in their quest for political clout in a process that once again is remaking the political and ecological landscape of Spain.

Conclusion: Recentring Scale and the Contested Politics of Rescaling

The production of spatial configurations as socioenvironmental cyborgs, part social part natural, excavated through the analysis of the circulation of hybridized water (water that is simultaneously physical and embodies deep sociocultural and political-economic meaning) opens up a new arena for thinking and acting. This arena is neither local nor global, but weaves a network that is always simultaneously deeply localized and extends its reach over certain scales, and certain spatial surfaces. The tensions, conflicts, and forces that flow with the water through the body, the city, the region, and the globe shape a continuously shifting power geometry, organized in a perpetually shifting and contested scalar configuration.

The examples illustrate how the production of socioecological scales is centred on the social transformation of nature and the construction of socioecological and political-ecological scalar gestalts. Concrete geographies, with choreographies of uneven and shifting social power relations, are etched into these ecological, social, political, or institutional scalar configurations. These processes are infused with contested and contestable strategies of individuals and social groups, who mobilize spatial scales as part of struggles for control and empowerment, and contest the power geometries of extant scalar gestalts. Needless to say, the mobilization of scale, the occupation of geographical scale, and the production of scale are central moments in such processes of sociospatial change. Struggling for the command of scale, or strategizing around excluding particular groups from the performative capabilities of certain scales, shapes social processes, defines relative empowerment and disempowerment, and gives rise to very specific sociospatial relations.

The politics of scale, then, although pivotally focused on the mobilization and appropriation of (metabolized) nature, necessitates a careful negotiation of the tensions, conflicts, and contradictions within and between scalar formations. Everyday bodily struggles for accessing water in Guayaquil's suburbs fuse with local politics, national economic processes, and international lending mechanisms, in ways that are often very contradictory and extremely difficult to negotiate or reconcile. Similarly, the up-scaling of Spanish water politics and engineering to the national scale mobilizes scalar politics that range from the reaffirmation of regionalist claims for autonomy, and demands from ecologists for a radical transformation of water practices, to the mobilization of the European Union as possible political ally or financial donor. Forging scalar alliances may be a tortuous and extremely difficult process, particularly for subaltern groups, for whom loyalty to and an insertion into a local social and physical ecology is of prime importance, and who are faced with the scalar mobilizations commanded by hegemonic global projects (such as global deregulation and free trade). The historical geography of capitalism is littered with examples of how sociospatial conflicts prevent the formation of "scaled" alliances, particularly by those that are already disempowered. Yet, a progressive politics of scale and the mobilization of scale are rapidly becoming key components in strategies to produce the democratic and inclusive social and ecological spaces that many of us dream of inhabiting.

Acknowledgements

This research was supported by the Belgian State who paid for the Ecuadorian research and the European Union who funded the Spanish research. I acknowledge this. I would particularly like to thank Eric Sheppard and Robert McMaster for their generous and detailed comments and suggestions on an earlier draft of this paper. I am also grateful to my current and past graduate students and colleagues, in particular Guy Baeten, Karen Bakker, Esteban Castro, Stuart Franklin, Maria Kaika, Ben Page, and Justus Uitermark, for the many discussions, arguments, suggestions, and critiques that helped shape these arguments. Of course, none of them should be held accountable for the final product and its remaining errors of fact or reasoning.

References

Aguirre, R. 1984: *Estado y Vivienda en Guayaquil*, Colecciòn Tesis, nr. 4, Facultad Latinoamericano de Ciencias Sociales (FLACSO), Quito.
Armstrong, W. and McGee, T. 1985: *Theatres of Accumulation*. London: Methuen.
Báez, R. 1985: Apogeo y Decadencia del Modelo Agroexportador – Periodo de la Segunda Postguerra. In R. Agoglia (ed.), *Historiografia Ecuatoriana*. Quito: Banco Central del Ecuador/Corporación Editora Nacional, 549–62.

Báez, R, 1992: Visión de la Economia Ecuatoriana, 1948–1970. In *Ecuador de la Postguerra – Estudios en Homenaje a Guillermo Perez Chiriboga*. Quito: Banco Central del Ecuador, 43–56.

Benton, T. (ed.) 1996: *The Greening of Marxism*. New York: Guilford Press.

Berndt, C. 2000: The rescaling of labour regulation in Germany: From national and regional corporatism to intrafirm welfare? *Environment and Planning A*, 32: 1569–92.

Bocco, A. 1987: *Auge Perolero, Modernizaciòn y Subdesarollo: El Ecuador de los Anos Setenta*. Quito: Coporaciòn Editora Nacional.

Bock, S. 1988: *Quito, Guayaquil: Identificaciòn Arquitectural y Evoluciòn SocioEconòmica en el Ecuador (1850–1987)*. Guayaquil: Instituto Francès de Estudios Andinos, Lima and Centro de Estudios Regionales de Guayaquil.

Boyle, M. 2000: Euro-regionalism and struggles over scales of governance: The politics of Ireland's regionalization approach to structural fund allocations 2000–2006. *Political Geography*, 19: 737–69.

Brenner N. 1997: State territorial restructuring and the production of spatial scale. *Political Geography*, 16: 273–306.

Brenner, N. 1998: Between fixity and motion: accumulation, territorial organization and the historical geography of spatial scales. *Environment and Planning D: Society and Space*, 16: 459–81.

Brenner, N. 1999: Globalization as reterritorialization: The re-scaling of urban governance in the European Union. *Urban Studies*, 36: 431–51.

Brunhes, J. 1920: *Geographie Humaine de la France*. Paris: Hanotaux.

Cano García, G. 1992: Confederaciones Hidrográficas. In A. Gil Olicna and A. Morales Gil (eds), *Hitos Históricos de los Regadíos Españoles*, Madrid: Ministerio de Agricultura. Pesca y Alimentación, 309–34.

Carrión, F. 1992: Evolución del Espacio Urbano Ecuatoriano', in E. Ayala (ed.), *Nueva Historia del Ecuador – Volumen 12 – Ensayos Generales I*. Quito: Corporación Editora Nacional/Editorial Grijalbo Ecuatoriana, 37–72.

Castree, N. 1995: The nature of produced nature: materiality and knowledge construction in marxism. *Antipode*, 27: 12–48.

Castree, N. 1999: Geographic scale and grass-roots internationalism: The Liverpool dock dispute, 1995–1998. *Economic Geography*, 75: 272–92.

Chiriboga, M. 1980: *Jornaleros y Gran Proprietarios en 135 Años de Exportacion Cacaotera (1790–1925)* Quito: Ed. CIESE – Consejo Provincial de Pichincha.

Chorley, R. J. 1969: The drainage basin as the fundamental geographic unit. In R. J. Chorley (ed.) *Introduction to Physical Hydrology*, London: Methuen, 37–59.

Collinge, C. 1999: Self-organization of society by scale: a spatial reworking of regulation theory. *Environment and Planning D: Society and Space*, 17: 557–74.

Cortez, P. Y. 1992: Lucha Sindical y Popular en un Periodo de Transición, In *El Ecuador de la Postguerra*. Quito: Banco Central del Ecuador, 543–69.

Cox, K. R. 1998: Spaces of dependence, spaces of engagement and the politics of scale, or: looking for local politics. *Political Geography*, 17: 1–23.

del Moral Ituarte, L. 1995: El Origen de la Organización Administrative del Agua y de los Estudios Hidrológicos en España. El Caso de la Cuence del Guadalquivir. *Estudios Geográficos*, 56: 371–93.

del Moral Ituarte, L. 1996: Sequía y Crisis de Sostenibilidad del Modelo de Gestión Hidráulica. In M. V. Marzol, P. Dorta and P. Valladares (eds) *Clima y Agua – La Gestión de un Recurso Climático*, Madrid: La Laguna, 179–87.

del Moral Ituarte, L. 1998: L'état de la Politique Hydraulique en Espagne. *Hérodote*, 91: 118–38.

Delaney, D. and Leitner, H. 1997: The political construction of scale. *Political Geography*, 16: 93–7.

Dicken, P., Kelly, P. F., Olds, K. and Wai-Chung Yeung, H. 2001: Chains and networks, territories and scales: Towards a relational framework for analysing the global economy. *Global Networks*, 1(2): 89–112.

Driever, S. L. 1998: "And since Heaven has Filled Spain with Goods and Gifts": Lucas Mallada, the regeneracionist movement, and the Spanish environment, 1881–90. *Journal of Historical Geography*, 24: 36–52.

Engels, F. 1844(1987): *The Condition Of The Working-Class In England*. London: Penguin.

Escobar. A. 2001: Culture sits in places: reflections on globalism and subaltern strategies of localization. *Political Geography*, 20: 139–74.

Farrell, G. 1989: *La Investigacion Economica en el Ecuador*. Antologia de las Ciencias Sociales, Nr. 4. Quito: ILDIS.

Giasante, C. 1999: In-depth analysis of relevant stakeholders: Guadalquivir river basin authority. Mimeographed paper, Department of Geography. University of Seville. 23 pp. (available from author)

Gómez Mendoza, J., and Ortega Cantero, N. 1987: Geografía y Regeneracionismo en España. *Sistema*, 7: 77–89.

Gómez Mendoza, J., and Ortega Cantero, N. 1992: Interplay of state and local concern in the management of natural resources: Hydraulics and forestry in Spain 1855–1936, *GeoJournal*, 26(2): 173–9.

Graigner, A. 1999: The role of spatial scale in sustainable development. *International Journal of Sustainable Development and World Ecology*, 6: 251–64.

Grijalva, W. M. 1990: La Economia Ecuatoriana de la Gran Recesión a la Crisis Bananero, In E. Ayala (ed.), *Nueva Historia del Ecuador*. Vol. 10. Quito: Corporación Editoral Nacional, 37–69.

Grundman, R. 1991: *Marxism and Ecology*. Oxford: Clarendon Press.

Guerrero, A. 1980: *Los Oligarcas del Cacao: Ensayo sobre la Accumulacion Originaria en el Ecuador*. Quito: Ed. El Conejo.

Haraway, D. 1991: *Simians, Cyborgs and Women – The Reinvention of Nature*. London: Free Association Books.

Haraway, D. 1997: *Modest-Witness@ Second-Millennium.FemaleMan© – Meets_ OncoMouse™*. London: Routledge.

Harvey, D. 1996: *Nature, Justice and the Geography of Difference*. Oxford: Blackwell.

Herod, A. 1991: The production of scale in United States labour relations. *Area*, 23: 82–8.

Herod, A. 2001: *Labor Goegraphies*. New York: Guilford Press.

Howitt, R. 1993: "A world in a grain of sand": Towards a reconceptualization of geographical scale. *Australian Geographer*, 24: 33–44.

Hughes J. 2000: *Ecology and Historical Materialism*. Cambridge: Cambridge University Press.

Hurtado, O. 1981: *El Poder Politico en el Ecuador*. Barcelona: Ed. Ariel.

Jonas, A. 1994: Editorial. *Environment and Planning D: Society and Space*, 12: 257–64.

Kelly, P. F. 1999: The geography and politics of globalization. *Progress in Human Geography*, 23: 379–400.

Larrea-Maldonado, A. 1982: Transnational companies and banana exports from Ecuador, 1948–72, *Northsouth*, 7(14): 3–42.

Lefebvre, H. 1991: *The Production of Space*. Oxford: Blackwell.

Leitner, H. 1997: Reconfiguring the spatiality of power: The construction of a supranational migration framework for the European Union. *Political Geography*, 16: 123–43.

León, P. 1992: Dos Decadas de Producción, Comercio Exterior y Reordenamiento del Espacio. In *Ecuador de la Postguerra – Estudios en Homenaje a Guillermo Perez Chiriboga*. Quito: Banco Central del Ecuador, 93–116.

Macías Picavea, R. 1899 (1977): *El Problema Nacional*. Madrid: Instituto de Estudios de Administración Local.

MacLeod, G. 1999: Place, politics and "scale dependence": exploring the structuration of Euro-regionalism. *European Urban and Regional Studies*, 6: 231–253.

MacLeod, G. and Goodwin, M. 1999: Reconstructing an urban and regional political economy: on state, politics, scale, and explanation. *Political Geography*, 18: 697–730.

Mallada, L. 1890(1969): *Los Males de la Patria y la Futura Revolución Espanola*. Madrid. Alianza Editorial.

Marston, S. 2000: The social construction of scale. *Progress in Human Geography*, 24: 219–42.

Massey, D. 1993. Power-geometry and a progressive sense of place. In J. Bird, B. Curtis, T. Putnam, G. Robertson and L. Tickner (eds) *Mapping the Futures: Local Cultures Global Change*, London: Routledge, 59–70.

Mateu Bellés, J. F. 1994: Planificación Hidráulica de las Divisiones Hidrológicas 1865–1899. Mimeographed paper, Department of Geography. University of Valencia (available from author).

Mateu Bellés, J. F. 1995: Planificación Hidráulica de las Divisiones Hidrológicas. In A. Gil Olicna and A. Morales Gil (eds) *Planificación Hidráulica en España*, Murcia: Fundación Caja del Mediterráneo, 69–106.

Merrifield, A. 1993: Place and space: A Lefebvrian reconciliation, *Transactions of the Institute of British Geographers*, N.S.18: 516–31.

Miller, B. 1997: Political action and the geography of defense investment: geographical scale and the representation of the Massachusetts miracle. *Political Geography*, 16: 171–85.

Rojas, M. and Villavicencio, G. 1988: *El Proceso Urbano de Guayaquil*. Quito/Guayaquil: Instituto Latinoamericano de Investigaciones Sociales y Centro de Estudios Regionales de Guayaquil.

Sadler, D. 2000: Organizing European labour: governance, production, trade unions and the question of scale. *Transactions of the Institute of British Geographers*. N.S.25, 135–52.

Schodt, D. W. 1987: *Ecuador: An Andean Enigma*. Boulder, CO: Westview.

Silvern, S. E. 1999: Scales of justice: Law, American Indian treaty rights and the political construction of scale. *Political Geography*, 18: 639–68.

Smith, N. 1984: *Uneven Development: Nature, Capital and the Production of Space*. Oxford: Blackwell.

Smith, N. 1993: Homeless/global: Scaling places. In J. Bird, B. Curtis, T. Putnam, G. Robertson and L. Tickner (eds), *Mapping the Futures: Local Cultures Global Change*, London: Routledge, 87–120.

Smith, N. 1996: The production of nature. In G. Robertson, M. Mash, L. Tickner, J. Bird, B. Curtis and T. Putnam (eds), *FutureNatural – Nature/Science/Culture*, London: Routledge, 35–54.

Smith, N. 1998. Antinomies of space and nature in Henri Lefebvre's *The Production of Space*. In A. Light and J. M. Smith (eds), *Philosophy and Geography II: The Production of Public Space*, London/New York: Rowman and Littlefield, 49–70.

Smith, N. and Dennis, W. 1987: The restructuring of geographical scale: coalescence and fragmentation of the northern core region. *Economic Geography*, 63: 160–82.

Smith, T. C. 1969: The drainage basin as an historical unit for human activity. In R. J. Chorley (ed), *Introduction to Geographical Hydrology*, London: Methuen, 20–29.

Swyngedouw, E. 1989: The heart of the place: The resurrection of locality in an age of hyperspace. *Geografiska Annaler*, 71B, 31–42.

Swyngedouw, E. 1992: Territorial organization and the space/technology nexus. *Transactions of the Institute of British Geographers*, n.s.17: 417–33.

Swyngedouw, E. 1995: The contradictions of urban water provision in Latin America. *Third World Planning Review*, 17: 387–405.

Swyngedouw, E. 1996a: The city as a hybrid – on nature, society and cyborg urbanization. *Capitalism, Nature, Socialism*, 7: 65–80.

Swyngedouw, E. 1996b: Reconstructing citizenship, the re-scaling of the state and the new authoritarianism: closing the Belgian mines. *Urban Studies*, 33: 1499–521.

Swyngedouw, E. 1997a: Neither global nor local: "Glocalization" and the politics of scale. In K. Cox (ed), *Spaces of Globalization: Reasserting the Power of the Local*, New York: Guilford, 137–66.

Swyngedouw, E. 1997b: Excluding the other: The contested production of a new "Gestalt of Scale" and the politics of marginalization. In R. Lee and J. Wills (eds), *Geographies of Economies*, London: Edward Arnold, 167–77.

Swyngedouw, E. 1997c: Power, nature and the city. the conquest of water and the political ecology of urbanization in Guayaquil, Ecuador: 1880–1980. *Environment and Planning A*, 29: 311–32.

Swyngedouw, E. 1998: Homing in and spacing out: Re-configuring scale. In H. Gebhardt, G. Heinritz, and R. Weissner (eds), *Europa im Globalisieringsprozess von Wirtschaft und Gesellschaft*, Stuttgart: Franz Steiner Verlag, 81–100.

Swyngedouw, E. 1999a: Marxism and historical-geographical materialism: A spectre is haunting geography. *Scottish Geographical Magazine*, 115(2): 91–102.

Swyngedouw, E. 1999b: Modernity and hibridity: Nature, *Regeneracionismo*, and the production of the Spanish waterscape, 1890–1930. *Annals of the Association of American Geographers*, 89: 443–65.

Swyngedouw, E. 2000a: Authoritarian governance, power and the politics of rescaling. *Environment and Planning D: Society and Space*, 18: 63–76.

Swyngedouw, E. 2000b. Elite power, global forces and the political economy of "glocal" development. In Clark, G., Feldman, M. and Gertler, M. (eds) *Handbook of Economic Geography*, Oxford: Oxford University Press, 541–558.

Tobar, L. 1992: Panoramo Agrario desde 1948. In *Ecuador de la Postguerra – Estudios en Homenaje a Guillermo Perez Chiriboga*. Quito: Banco Central del Ecuador, 227–61.

Towers, G. 2000: Applying the political geography of scale: grassroots strategies and environmental justice. *The Professional Geographer*, 52: 23–36.

Trujillo, J. 1992: Expansión de la Frontera Agricola. In *El Ecuador de la Postguerra– Estudios en Homenaje a Guillermo Perez Chiriboga*. Quito: Banco Central del Ecuador, 201–26.

Villanueva Larraya, G. 1991: "*La Politica Hidráulica" durante la Restauración 1874–1923*. Madrid: Universidad Nacional de Educación a Distancia.

Walsh, J. 2000: Organizing the scale of labor regulation in the United States: Service-sector activism in the city. *Environment and Planning A*, 32, 1593–1610.

Zeller, C. 2000: Rescaling power relations between trade unions and corporate management in a globalising pharmaceutical industry: The case of the acquisition of Boehringer Mannheim by Hoffman-La Roche. *Environment and Planning A*, 32, 1545–67.

Zimmerer, K. S. 2000: Rescaling irrigation in Latin America: The cultural images and political ecology of water resources. *Ecumene*, 7: 150–75.

7 Scales of Cybergeography

Michael F. Goodchild

As other chapters in this book will have already made clear, the word *scale* has many meanings – it is highly *overloaded*. Geographers are of course most interested in *spatial* scale, and less interested in scales of variation in time, or in other dimensions. In this chapter, I explore the significance of scale in the context of what I term the *digital transition*, or the conspicuous take-up of digital technology in all aspects of society – the workplace, the arts, entertainment, and education.

Much has been written in recent years about the digital transition (for discussions with a geographic focus see for example Dodge and Kitchin, 2001; Janelle and Hodge, 2000; Leinbach and Brunn, 2001). We read about the *new economy*, and associated ideas of trading in information rather than tangible goods; a shift from customary patterns of shopping to on-line electronic commerce; a growing disparity in wage rates between those who are able to contribute skills and those who are not; and even the now-discredited speculation that the new economy is somehow free from the traditional business cycle. The information age is said to make it possible for more people to work from home; is widely believed to be more environmentally responsible than the industrial age; and is often said to be conducive to a better-informed citizenry and a higher quality of political debate. However, many of these optimistic assertions may be false; and to date the new economy has clearly not been egalitarian, but rather has tended to exacerbate existing cleavages, between rich and poor communities, and between developed and underdeveloped countries. Despite vast increases in the flow of information, the growing digital divide remains one of the most discouraging aspects of the past decade.

The reasons for the digital transition are not hard to find. Although the term *digital* is preferred, the reference to fingers is clearly misplaced, because digital technology is based on a binary counting system, not a decimal one. Virtually all forms of human communication, with the obvious exception of direct communication by voice or gesture, now pass through a digital form at some point. Standard coding systems have emerged to support the representation of numbers, text, graphics and images, music, and even maps in digital form. Digital encoding is attractive because it can be accompanied by simple procedures that identify and remove errors; and because massive economies of scale can be obtained if a single technology can support the transmission of virtually any kind of information. For example, the devices that handle the packets of bits that travel through the Internet are in no way dependent on the specific meaning of the packets to their senders and receivers, since those meanings may range from pieces of music downloaded by high school students to encrypted, top-secret military communications – both are simply strings of zeroes and ones. In this way the binary alphabet achieves far greater scale economies than were possible with earlier coding systems, such as the Morse code of the telegraph, which was never able to encode music efficiently.

The chapter is structured as a series of four variations on a common theme, the effects of the digital transition on scale. First, digital technology is providing new flexibility in the ways in which knowledge of the Earth's surface is represented. Scale is an important property of any representation, but the ways in which the property is measured are strongly affected by the digital transition. I explore these impacts, and propose measures that are more appropriate for digital representations. Second, the digital transition is enabling sharing of information to an unprecedented degree. Geographic information is voluminous, and I show that it can easily swamp communication networks. But arguments based on scale can be used to show that such concerns are largely unfounded. Third, the digital transition and associated developments in geographic information technology have greatly changed the economics of the production and dissemination of geographic information, which is in the process of being reorganized at novel scales. Finally, it is often argued that the digital transition is reducing or even removing the importance of space in human organization. If this is true, the importance of scale must also diminish. I explore the arguments for and against this proposition.

Scale in Digital Representations

Given the profound change represented by the digital transition, it is worth asking whether there are instances of concepts that fail to survive – that were prominent in the earlier era, but are now losing their meaning and therefore their use. Before the recent growth of technology based on the binary alphabet, one of the primary means of communicating geographic information was the paper map. This is an instance of *analog* representation, since the world is represented not in a digital code, but as a proportional model. Analog representations are the basis of the Bell telephone (electrical signals proportional to changes of atmospheric pressure, now largely replaced with signals encoded in binary form), traditional photography, and many other communication media. The constant of proportionality is an essential parameter of any analog representation, and in the case of paper maps (and many other analog representations) the term *scale* is used. *Representative fraction* is a more intuitively appropriate term, because it immediately suggests a ratio between the representative model and the real world. But a digital representation is not proportional, and the representative fraction is not meaningful for digital geographic databases – there is no distance in the representation that can be compared to distance on the ground.

Goodchild and Proctor (1997) argue that the representative fraction has nevertheless persisted through the digital transition, because of its familiarity – an instance of a more general pattern in which concepts associated with an earlier technology persist into a new era (the term *horseless carriage* is often cited, because it defines the new concept – the automobile – in terms of earlier ones). In order for this to be possible, a complex system of conventions has arisen that allows representative fractions to be assigned to databases that do not intuitively have them. The most obvious of these is the convention adopted for databases that originated as paper maps (and were transformed by digitizing). In this case the representative fraction (RF) of the digital version is by convention the RF of the source analog document. It is implicit in this convention that the act of digitizing does not significantly corrupt (or improve) the contents, through omissions or the introduction of positional errors, and thus in effect degrade the RF.

RF is a particularly successful scale parameter of paper maps, because it acts as an effective surrogate for many other properties. Mapping agencies have adopted practices that link the collection of features shown on maps to the RF (e.g., a US topographic map at 1:24,000 shows major streets but not buildings), allowing the RF to act as a surrogate for content. RF also acts as a surrogate for positional accuracy, a relationship that is established through national map accuracy standards which link them directly. For example, the US National Map Accuracy Standard (USGS, 1999) specifies that on maps with an RF of 1:20,000 or larger (more detailed) 90 percent of points should be within 1/30th inch of their true positions; for maps at coarser scales, the standard is 1/50th inch (e.g., for maps of 1:24,000 the standard corresponds to an error on the ground of 12.2 m). Thus if positional accuracy is known for a digital database, it is possible by convention to specify an RF, and this convention is often used in the case of digital orthophotos (a positional accuracy of 6 m is linked via the national map accuracy standard to an RF of 1:12,000, because 6 m is the approximate specified accuracy for maps of that RF). But for digital databases there need not be any such precise relationship between positional accuracy, spatial resolution, and content.

RF fails to survive the digital transition except by convention, raising the question of how positional accuracy, spatial resolution, content, and other expressions of level of detail should be defined for digital databases. In the case of raster data sets, spatial resolution is readily defined by the cell size, since this is the minimum distance over which change is recorded. Define the linear measure of cell size (cell width) as S, an index that is useful for both analog and digital representations. Unfortunately vector representations do not provide a similarly rigorous way of defining S, and confuse this property of the representation with properties of the phenomenon (Figure 7.1). For example, consider an isotherm map used in climatology. Its spatial resolution is related to the spacing of weather stations at which atmospheric temperature is recorded, but there is no obvious way of determining S from such a map, either from the observed spatial variation in temperature, or from the spacing of weather stations. It may be preferable to treat S as varying spatially, with greater spatial detail in areas where weather stations are closer together. Urban mapping presents similar problems; for example, census tracts are designed to have roughly equal populations, and thus vary in size, being typically smallest in high-density urban cores. But it may also make sense that they are larger in suburban areas because spatial variation there is less rapid, and perhaps also because the attenuating effects of distance on social interactions are less in which case variable spatial resolution may have rigorous justification.

Besides level of detail, the term *scale* is also used in the sense of extent, or scope. Define L as a linear dimension of extent, equal to the diameter of a circular project area, or the edge length of a square project area, or the square

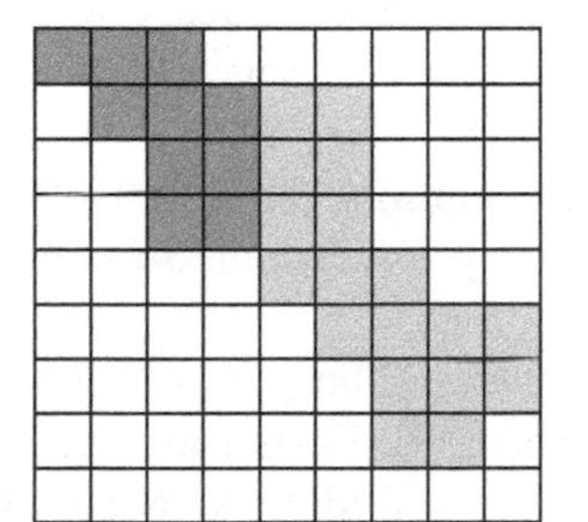

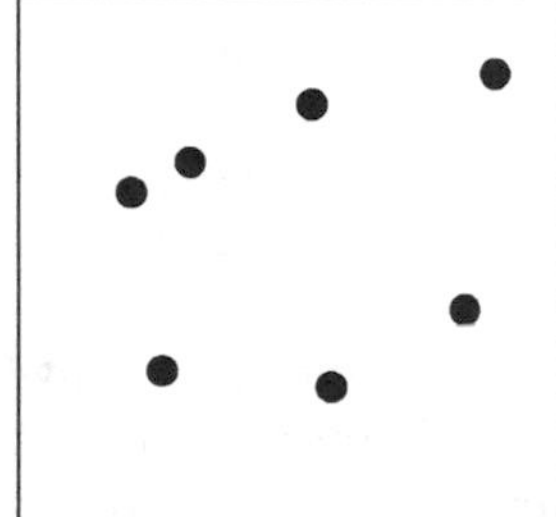

Figure 7.1 *Spatial resolution (S) is well-defined for a raster data set (left) as the square root of cell area. But the irregular spacing of sample points in a vector data set (right) implies that spatial resolution is not uniform across the sampled area.*

root of area for an oddly shaped area. A project covering the entire surface of the Earth has an *L* of approximately 23,600 km (square root of surface area), and the parameter is readily defined for topographic quadrangles, Landsat scenes, etc., whether the data are in analog or digital form.

Dimensionless constants are often useful in science, because their values are invariant under changes in units of measurement. Thus the Reynold's Number is one of a number of such dimensionless parameters widely used in hydrology. The ratio *L/S* is dimensionless, and equal to the ratio of two scales – the *large* and *small* meanings of the term – or the ratio of extent to resolution, both properties that unlike RF survive the digital transition.

Although *L/S* can in principle take any value greater than or equal to 1, in practice its range appears to be remarkably limited. The display screens of modern personal computers are typically in the *megapel*, or million-pixel range, a common configuration having 1024 columns and 780 rows, or an *L/S* ratio of order 10^3. Although some devices such as digital cameras have higher values, it appears that there is little pressure in the marketplace to increase this value – that most people find the resolution of computer displays adequate for most purposes. There may be a physiological explanation for this, since the *L/S* ratio of the human eye (ratio of the spatial extent of the retina to the diameter of a retinal cell) is of order 10^4, and the computer screen is not expected to occupy the entire visual field. Thus any greater resolution would have little perceptual value. If we take the spatial resolution of a paper map to be defined by the width of a pen, or the size of the smallest symbol that can readily be discerned by the map user, then its *L/S* ratio is also of order 10^3. A Landsat scene has an *L/S* ratio of approximately 3,000, and other sources of imagery tile their products into similar multiples of picture elements. In summary, although both *L* and *S* vary over many orders of magnitude in geographic data (in the case of *L*, from the 20,000 km of a whole Earth to the 200 m of a detailed biogeographical study), the ratio *L/S* appears to be remarkably persistent in practice (Table 7.1). The reasons appear to be both technical and cognitive. On the technical side, and particularly for raster data sets, *L/S* is a predictor of data volume, which tends to rise as the square of the ratio, eventually swamping current technology. On the cognitive side, the relationship between *L/S* is related to the eye's sensory limitations and the processing abilities of the eye/brain.

It is interesting to speculate on whether similar patterns can be found in other forms of information. It is rare to find journal articles with lengths outside the range 5,000 to 10,000 words, and it is rare to find bound volumes with lengths outside the range 50 to 2,000 pages. But in neither case is there an obvious equivalent of resolution, and in both cases there are many reasons, some technical and some behavioral, for limits to extent.

Table 7.1 *Comparison of the approximate extent and resolution of some instances of geographical information, and of the human visual system*

Example	*Extent (L) definition*	*Extent (L) estimate*	*Resolution (S) definition*	*Resolution (S) estimate*	*L/S ratio*
Computer screen	Square root of screen area	250 mm	Square root of pixel area	0.25 mm	1,000
Landsat Thematic Mapper scene	Square root of scene area	90 km	Square root of pixel area	30.0 m	3,000
Paper map	Square root of map sheet area	1 m	Typical pen width	0.5 mm	2,000
Human retina	Square root of retinal area	25 mm	Square root of retinal cell area	0.0025 mm	10,000

Scale in the Sharing of Geographic Information

In this section I relate these arguments to the *scaling properties* of geographic information. The verb *to scale* is frequently used in discussions of information technology, and refers to the behavior of technology under changes in the volumes of information being processed. A technology is said to scale if its performance fails to degrade substantially when volume increases. I first examine the scale of textual information, since textual media still largely dominate scholarly communication, and then move to examine visual and geographic information.

Human speech achieves rates of transmission of order 10^5 characters or bytes per hour. A typical book contains order 10^6 characters or bytes, and a prolific academic might write order 10^8 bytes in a lifetime (easily stored on a removable diskette). An average reader processes order 10^5 bytes per hour, and perhaps order 10^9 bytes, or the contents of order 10^3 average books, in an average reading lifetime. A major research library of order 10^6 books contains order 10^{12} bytes of text information, or 1 terabyte. The fact that such a volume of information can now be stored on a device no larger than a piece of furniture, and accessed simultaneously and independently by thousands of readers, is one of the motivating forces behind current research into digital libraries (Arms, 2000).

Visual information is in general much more voluminous than textual information when represented in digital form, and the limited amount of visual information present in research libraries, in the form of images or maps, can easily overwhelm these estimates. A single Landsat scene contains order 10^8 bytes, depending on the degree of compression, while the replay of a DVD video requires display of order 10^9 bytes per hour. If we take video replay as a crude basis for estimating the processing capabilities of the human vision system, it appears that we process order 10^{13} bytes per year, or order 10^{15} bytes (one petabyte) in a lifetime.

Finally, consider geographic information. The EOS series of satellites is expected to provide order 10^{12} bytes per day of information. The Earth's surface has order 10^{15} sq m, so a complete coverage at this resolution with one byte per sq m would produce order 10^{15} bytes, before compression. A person viewing such data at a rate of 1 sq km per second, by filling a computer screen every second with 10^6 fresh picture elements, would take approximately a working lifetime to view the entire surface of the Earth.

It is interesting to compare these estimates with statistics for the Internet. The network includes order 10^7 servers, and order 10^8 users. Estimates of the total volume of information accessible via the Internet are notoriously unreliable, but generally exceed order 10^{15} bytes, of which only a fraction are catalogued by the Internet search engines. Bandwidths vary, but take 10^8 bytes/sec as the capacity of a "fat" pipe. Then if all Internet users required access to all information, order 10^{23} bytes would have to be transmitted, requiring order 10^{15} seconds through our fat pipe, or order 10^8 years. Clearly the Internet is capable of massive congestion.

But now recall the result of the previous section, that in practice the *L/S* ratios exhibited by geographic data sets fall into a fairly narrow range. Although in principle the exchange of such data could quickly swamp the Internet, in practice a user interested in studying the entire Earth's surface ($L = 20,000$ km) is unlikely to expect or demand a spatial resolution much finer than 20 km ($L/S = 10^3$), and conversely a study requiring a spatial resolution of 1 m is unlikely to extend over much more than 1 km. Although volume rises as $(L/S)^2$, such users can be readily accommodated by the bandwidths and storage devices available in most research institutions.

Scale in the Production of Geographic Information

Traditional methods for the production of geographic information in the form of maps have been highly capital-intensive, requiring the fielding of large teams of surveyors; the acquisition of photography from airborne cameras; massive photogrammetric equipment; and high-resolution color printing systems. Many countries, particularly those with large land areas, have regarded geographic information as a public good, produced by agencies of national governments, often under military control because of the strategic and tactical value of maps. Production of geographic information has thus been centralized at the national scale.

In recent years rapid advances in geographic information technology have led to vast and fundamental changes in the economics of production. These include the development of "soft" or computerized photogrammetry, which allows massive mechanical systems to be replaced by simple computer applications; and the Global Positioning System, which measures position on the Earth's surface cheaply and directly, and to sufficient accuracy for many types of geographic information. Images from satellites are in many cases sufficiently detailed to replace costly aerial photography. Finally, the digital transition in geographic information has meant that investment in printing is no longer necessary, since the contents of maps can now be disseminated rapidly and essentially without cost.

These changes have greatly reduced the high fixed costs associated with centralized geographic information production, and effectively removed the need for massive economies of scale. Today, it is possible for a local agency, a small corporation, or even an individual farmer (Longley et al., 2001;

NRC, 1997) to collect, interpret, and analyze geographic information, and to do so at costs that are low enough to justify the necessary investment and expense. In effect, the production of geographic information has shifted from a centralized system organized at a national scale, to a dispersed network operating at a wide range of scales that correspond to the domains of interest of national, state, or local governments, corporations, or individuals.

The Digital Transition and the Breakdown of Spatial Organization

Geographers have long sought explanations for the patterns of spatial differentiation that are observed on the Earth's surface, and have developed theories that attribute them in part to the costs of overcoming separation. Transport costs, for example, are held to be major determinants of industrial location, and travel costs are similarly determinants of market location. Moreover the scales exhibited by human organization on the Earth's surface directly reflect such determinants. The *range* parameter of the central place theories of Christaller (1966) and Lösch (1954) is defined as the distance consumers are willing to travel to obtain a good, while the *threshold* parameter is defined as the number of consumers needed to support the offering of the good. Together, range and threshold define the spacing of settlements offering the good. If the marginal costs of overcoming separation are driven to zero, and with them a major factor explaining spatial differentiation, it is reasonable to ask what the effects might be on the scales of organization of the associated human activities.

The Internet achieves a close-to-zero marginal cost of distance not through the low costs of laying fiber, which are high, but through the enormous economies of scale that are possible because of the massive capacity of fiber links (some of the major backbone links in the US network now have capacities of over 2 gigabits per second, equivalent to 2 million average email messages of 100 characters per second). In addition, there are few mechanisms for passing any marginal costs on to the consumer, and thus affecting consumer behavior, because the Internet has evolved as a service that is essentially free to its users – few users pay the costs associated with sending packets, whatever the distance over which the packet is sent.

The breakdown of spatial organization that accompanies a decreased role of distance is easy to visualize. Consider a large array of picture elements, such as a remotely sensed scene. Select pairs of picture elements at random, without respect to location, and swap their contents. Spatial patterns begin to disappear, and after an appropriate number of swaps it is difficult to detect any residue of the original image (Figure 7.2). A screen filled with picture elements of random colors would be meaningless, and would be perceived by the eye as a uniform gray. In essence, much spatial organization results from connections between metrics of space, such as distance, and the behaviors of actors. When those connections are broken, a major cause of spatial organization is removed.

Although the Internet reduces the marginal cost of transporting information to close to zero, nevertheless all human activities must be located somewhere, including the electronic bits that represent information. Transport of commodities other than information still incurs marginal costs, and thus continues to result in scale effects. Other activities such as mining and agriculture are tied to properties of places, and display spatial patterns and scales that derive from the spatial distributions of the underlying resources. But with respect to information, spatial metrics are now essentially irrelevant – there are no longer real costs to the communicator associated with distance, no significant time delays, and no significant bandwidth constraints that are distance-related.

Consider a complex biological organism, such as the human body. Its central nervous system allows messages received by the brain to be associated with specific locations – we sense where pain signals originate. However, the immune system has functional rather than spatial organization; an infection in one place must be fought by dispatching large numbers of white blood cells throughout the system, only a small fraction of which will actually be needed to fight a localized infection. In the very early stages of fetal development cells are somehow capable of sensing their spatial locations as they differentiate, but these interactions extend only over a few cells, and an adult cell in the foot has no way of knowing or responding to its distance from a cell in the head.

Complex biological systems thus exhibit aspects both of *spatial* organization at different scales, and of *functional* organization. The Internet similarly

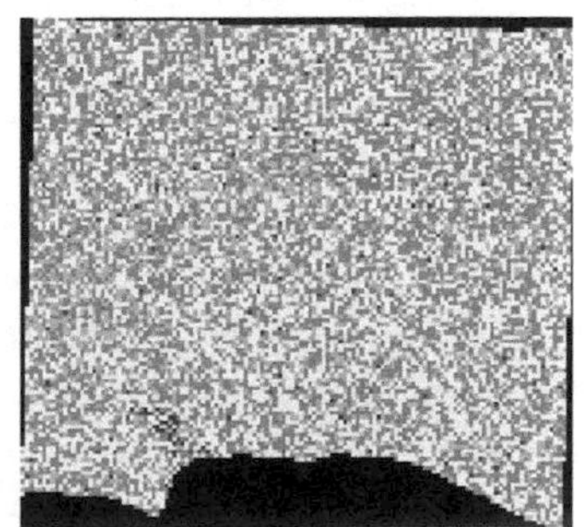

Figure 7.2 *Spatial elements can begin to disappear. The spatial organization inherent in the left-hand image (classified land-cover in part of Goleta, CA) can be destroyed by randomly swapping the contents of pairs of pixels (right).*

exhibits aspects of spatial organization in the processes by which a packet is directed from a sender to a receiver. From the user's perspective, however, there is very little evidence of spatial organization at any scale. The earliest domains, such as .edu, differentiated on the basis of function. National domains are widely used outside the US (and some US state governments use both state and national domains, as in .nc.us), but a proposal to build a consistent geographically based system of addresses under a new .geo domain failed to be adopted by the Internet domain authority ICANN (International Corporation for the Assignment of Names and Numbers) in 2000 (see *http://www.sri.com/news/releases/10-23-00.html*). While earlier systems of communication such as the postal service and wired phone service have strong spatial organization (postal codes and telephone area codes are predominantly spatial), the Internet has clearly moved sharply in the direction of functional organization.

Arguments for a continued importance of spatial organization

Nevertheless, several arguments support the continued presence of a degree of spatial organization in the sharing of information on the Internet, and in related activities; and in turn support the persistence of scale effects in related activities. First, although the marginal cost of transporting information has been driven almost to zero, certain limitations remain that are significant in some circumstances. Cost is only one form of potentially distance-related impedance, and several others are important to users of the Internet. People further apart are more likely to speak different languages, and to experience related difficulties of communication. *Latency* measures the delays associated with transmitting packets between points, and although it is typically distance-related, most packets travel at close to the velocity of light. But latency is also related to the degree of congestion due to excessive traffic, which can be distance-related. Murnion and Healey (1998) have analyzed patterns of latency on a global basis, and shown weak correlations with distance that are in part due to the crossing of international boundaries. *Reliability* is also distance-related, since the potential for service interruption clearly increases with the distance separating the sender and receiver, and with the number of nationally based telecommunication networks encountered en route. Large web sites are sometimes *mirrored* at a very coarse, intercontinental scale, particularly when the user base is strongly multinational, because of specific bandwidth limitations. There have been periods, for example, when bandwidth between the US and the UK was significantly impacted during certain hours of the day, notably when US and UK working hours overlapped in the UK early afternoon. Such effects are capable of introducing coarse-scale differentiations of activities related to information exchange.

Although fiber-optic links achieve massive economies of scale between their endpoints, the Internet is far from uniformly accessible across space. The so-called "last mile" problem refers to the high cost of providing the low-volume links between high-volume endpoints and distributed business and residential customers, and their high marginal cost of distance. Wireless technology has the potential to reduce this marginal component, but remains largely unimplemented, with limited geographic coverage. Let $c(\boldsymbol{x})$ represent the cost of connecting to the Internet from location $\boldsymbol{x}$, a function that shows extreme local variability over short distances; neighboring houses on the same street can differ by as much as a factor of 1,000 in communication bandwidth. The fixed cost of two people at $\boldsymbol{x}_1$ and $\boldsymbol{x}_2$ connecting and communicating, $c(\boldsymbol{x}_1) + c(\boldsymbol{x}_2)$, is largely independent of the total distance between $\boldsymbol{x}_1$ and $\boldsymbol{x}_2$, but highly dependent on the local spatial variation of c. Intense local variation in access can lead in turn to intense local variation in activities that involve information exchange, such as the prevalence of telecommuting. Internet access retains a spatial organization, but of a particularly complex and fine-scaled kind.

Second, many patterns of Internet interaction reflect acquaintance networks that were established through physical contact, or require physical contact. For example, people who met in high school may continue their interaction by email. In a related example, Wheeler and O'Kelly (1998) have shown that Internet traffic from the Ohio State University (OSU) servers is strongly distance-related at the scale of the state, because it is determined to some extent by the locations of students, most of whom are from the state and attend OSU in person. Telecommuting patterns will be strongly influenced by distance if workers are required to spend part of their time at the employer's site. In all of these examples the pattern of information flows is linked to a pattern of physical flows at regional scales that incur distance-related costs.

Finally, much Internet interaction is driven by activities that are tied to physical locations and affected by transport costs. For example, locations of retail stores remain largely tied to local-scale residential distributions of consumers, and the information services that they require are similarly tied.

Storage of geographically referenced information

Many of these arguments can be exemplified by the case of geographically referenced information, or information that is associated with a particular *footprint* on the Earth's surface. Such information includes not only the contents of maps, but also reports, photographs, books, and even musical compositions associated with particular places. In this section I consider the question of where such information is likely to be found following the digital transition (see Goodchild, 1997, for a more detailed analysis of this problem).

Prior to the digital transition geographically referenced information was associated with specific physical media (bound volumes, map sheets, photographic film). Because the use of such media required physical access, it was necessary for each user either to acquire a copy, or to travel to a central facility such as a library. The scale of organization of libraries was dictated by the spatial distribution of users, and by the ranges and thresholds associated with the library's services.

In principle, the Internet allows a user located anywhere to access information stored anywhere, at costs that are unrelated to distance. Spatial organization in library services becomes a thing of the past, and it is no longer necessary for libraries to duplicate each others' contents. A user searching for a particular item of information might find it anywhere; and would have no way to predict where the information would be found, and no rational basis for deciding where to search.

However, geographically referenced information has properties that argue strongly for spatial rather than functional organization, and for persistence of scale as an organizing principle. First, interest in geographically referenced information is often related to distance: a street map of Paris tends to be of much more interest to a user located in Paris than to a user located in Los Angeles – unless the latter is planning a trip to Paris, for example. Second, in discussions of national spatial data policy (e.g., NRC, 1993) it is often argued that such information is best maintained locally, because of easy access to ground truth. Thus a street map of Paris is more likely to be up to date if produced and maintained by a Parisian agency, than if produced and maintained by an agency in Los Angeles. Third, in the previous section I argued that changes in the economics of data production have enabled cheap, rapid, and local production by much smaller groups and even by individuals.

These three arguments together suggest that after the digital transition geographically referenced information is most likely to be found physically located near the information's footprint; and that this process operates over scales comparable to the extent of the footprint (e.g., internationally for data sets with national footprints). For example, an analysis of the more than 1 million such information items in the Alexandria Digital Library, a major online collection of geographically referenced information, shows that the abundance of items declines globally with distance from the library's physical location at the University of California, Santa Barbara (Goodchild and Zhou, in press).

Concluding Comments

If distance is now irrelevant to many aspects of human organization (Cairncross, 1997), particularly those that are mediated by electronic communication, then the future human geography of the planet's surface will be vastly different. There are many technical and behavioral reasons to believe that distance will continue to be important to many human activities, as I argued in the third section of this chapter. However, the *parameters* of those distance effects will almost certainly be different, and consequently the scales associated with the organization of the associated activities will also be different.

The chapter has concentrated on scale in two distinct senses: as a measure associated with a representation, and as a parameter associated with a process. In principle the two senses are intimately linked, since it is impossible to understand a process by examining a representation unless the scale of the representation is at least as detailed as the scale of the process. For example, understanding the journey to work as a spatial process clearly requires much more detailed data than understanding international migration as a spatial process. But in practice the acquisition of data often proceeds somewhat independently of the investigation of process, and judgments often have to be made about the fitness of a data set for a particular use, based on metrics of scale. The first section of the chapter proposed metrics that are more appropriate for digital geographic information than the traditional representative fraction.

The dimensionless ratio of extent to resolution, or *L/S*, appears to be approximately invariant across a wide range of types of geographic information and forms of representation, for a combination of technical and cognitive reasons. This suggests practical constraints on the usefulness of large volumes of geographic information, and the possibility of design principles that can be used to anticipate data needs across a wide range of applications.

In the third and fourth sections of the chapter I explored the links between scale and process, with respect to activities that are impacted by the digital transition. The process of restructuring the production of geographic information is already well under way, and novel scales are already emerging. Library services are also being restructured, and offered at new scales. The specific example of geographically referenced information underscored the factors that appear likely to sustain a substantial degree of spatial organization in many human activities despite the pervasive digital transition. In short, distance is not disappearing as a basis for human organization; but the parameters that define its importance are changing, and new scales of organization are emerging as a consequence.

Acknowledgment

The Center for Spatially Integrated Social Science and the Alexandria Digital Library are supported by the US National Science Foundation.

References

Arms, W. Y. 2000: *Digital Libraries*. Cambridge, MA: MIT Press.

Cairncross, F. 1997: *The Death of Distance: How the Communications Revolution Will Change Our Lives*. Boston, MA: Harvard Business School Press.

Christaller, W. 1966: *Central Places in Southern Germany*. Translated by C.W. Baskin. Englewood Cliffs, NJ: Prentice Hall.

Dodge, M. and Kitchin, R. 2001: *Mapping Cyberspace*. London: Routledge.

Goodchild, M. F. 1997: Towards a geography of geographic information in a digital world. *Computers, Environment and Urban Systems*, 21: 377–91.

Goodchild, M. F. and Proctor, J. 1997: Scale in a digital geographic world. *Geographical and Environmental Modelling*, 1: 5–23.

Goodchild, M. F. and Zhou, J. 2003: Finding geographic information: collection-level metadata. *GeoInformatica*, 7: 95–112.

Janelle, D. G. and Hodge, D. C. (eds) 2000: *Information, Place, and Cyberspace*. New York: Springer.

Leinbach, T. R. and Brunn, S. D. (eds) 2001: *Worlds of E-Commerce*. New York: Wiley.

Longley, P. A., Goodchild, M. F., Maguire, D. J. and Rhind, D. W. (eds) 2001: *Geographic Information Systems and Science*. New York: Wiley.

Lösch, A. 1954: *The Economics of Location*. Translated by W. H. Woglom. New Haven, CT: Yale University Press.

Murnion, S. and Healey, R. G. 1998: Modeling distance decay effects in Web server information flows. *Geographical Analysis*, 30: 285–303.

National Research Council 1993: *Toward a Coordinated Spatial Data Infrastructure for the Nation*. Washington, DC: National Academy Press.

National Research Council 1997: *The Future of Spatial Data and Society*. Washington, DC: National Academy Press.

US Geological Survey 1999: *Map Accuracy Standards*. Fact Sheet FS–171–99. Washington, DC: US Geological Survey (http://mac.usgs. gov/mac/isb/pubs/factsheets/fs17199.html).

Wheeler, D. and O'Kelly, M. E. 1998: A method for generating cyber trade areas: a case study of The Ohio State University. Presentation at 1998 Annual Meeting, East Lakes Division, Association of American Geographers (October 31).

8 A Long Way from Home: Domesticating the Social Production of Scale

Sallie Marston

> Scale becomes the arena and the moment, both discursively and materially, where sociospatial power relations are contested and compromises are negotiated and regulated. (Swyngedouw, 1997: 140)

The attention paid by geographers to understanding the ways the production of scale is implicated in the production of space has been increasing dramatically over the last ten years. Remarkably, geographic scale has attracted attention across the discipline – with physical, human, and methodological geographers – also spilling out into related social sciences including sociology, anthropology, and political science. Indeed, this book is testimony to just how central scale theorizing has become for geographers in a fairly short time. For human geographers, especially critical human geographers, the focus on scale revolves around a recognition that it is socially produced through material and discursive practices. Theorization about scale attempts to understand how it is constructed or transformed within a particular historical and geographical context, and what implications particular manifestations of scale might have for social, cultural, political, and economic practices.

I have argued recently (Marston, 2000; Marston and Smith, 2001) that questions currently driving the scholarship on scale overwhelmingly situate capitalist production as their central concern while theoretically omitting the ways in which social reproduction is also implicated in scale construction processes. In this chapter, I build upon that argument by developing more fully the empirical case on which it rests; an extended examination of turn of the nineteenth-century scale production that in large part was the result of interactions between situated agents and the state. Many of the most powerful of these social agents operated through urban US female, white, and middle-class identity formations around issues of social reproduction as they became articulated with the growth of cities (and suburbs). I wish to pursue two related aims. The first is to show how a particular ensemble of social agents, bearing culturally significant subjectivities, seized a particular historical moment through their comprehension of a particular geographical arena, and influenced substantially, if not actually helped to determine, a particular scalar fix of state, territorial, and functional organization that further shaped the structure of the economy. I argue that the state, in producing scale or in any of its other practices, is the outcome of political mobilizations of a whole range of social agents, some of them surprisingly efficacious notwithstanding obstacles to their independent political agency. The state must ultimately favor the stability of the capitalist system, but the needs or demands of capital and labor (or significant fractions of each) must also be mediated. Thus, the cases I treat are meant to demonstrate not only why a particular scalar fix enables capital accumulation at a particular moment, but also how socially and culturally inscribed social agents, struggling over ideology and meaning systems, have interacted with the state to formulate that particular scalar fix. To put it rather bluntly, my first interest is to understand more fully the *social* production of scale by literally *embodying* the concrete forms of social action that are central to it.

My second interest is to use the historical cases to illustrate how our developing theory of scale production will be enhanced if we can gain a better grasp of the totality of the political economy of contemporary capitalism – especially the ways in which social reproduction is tied up with production (particularly through the relationship between capital and all forms of labor). Scale production is unquestionably a means for enabling capital accumulation; and surely capital accumulation rests on economic production as well as social reproduction, even though one may be more salient to accumulation than the other in any particular time or place. Thus, while state supported social reproduction may have declined in importance relative to contemporary state organization in the United States, this has not always been the case. In short, a particular scalar fix is only a "temporary spatialization of certain social assumptions" as Neil Smith (this volume) also contends.

In the 1970s, interest in the "urban question" reflected a moment when social reproduction – in the form of collective consumption provision – created a crisis for capital accumulation (Castells, 1977). This crisis was eventually resolved by the state's general retreat from the provision of these services. While a new scalar fix has emerged, changing the meaning and salience of social reproduction, the relevance of social reproduction to state formation has not disappeared entirely and can certainly reappear as a new contradiction or crisis or even as a solution. In sum, my second interest is to remind us that social reproduction deserves serious consideration in our theorizing of scale even though it is not especially central to contemporary state reterritorialization. I begin this chapter with a highly condensed review of approaches to scale that center on economic production. I then outline the theoretical literature on social reproduction that informs my empirical cases, highlighting the important structural forces that shaped the US city and urban culture, especially as they relate to the emergence and gendering of the middle class at the turn of the nineteenth century. These important forces include the production and deployment of domestic technologies; the consolidation of the local state as a fully public – as opposed to a quasi-private-enterprise increasingly oriented around public provision; and the assumption of social welfare provision by the federal state. In the third section I describe the scale construction that occurred around several women's groups operating at the local, regional, and national level. My aim here is to show that the turn-of-the-century-urban/suburban home/household – a key site of social reproduction in modern capitalism – became a site of political identity formation for middle-class white US women who used it to influence and configure state functions. This prolonged and well-documented moment contributed to a politics of scale that addressed and shaped, both materially and discursively, local, national and, to some extent, international capitalist political economies.

The Social Construction of Scale

Although human geographers have been interested in scale for over two decades (Taylor, 1982; Kirby, 1985; Holly, 1978), the novelty in recent thinking is the commitment to recognizing that scale is made, and not an ontologically given category; that scale is not a preordained hierarchical nomenclature for ordering the world, but rather a contingent outcome of the tensions between structural forces and the interventions of human agents. This apparently simple point is critical because recognizing that scale is not infinitely fixed forces us to understand and theorize the process whereby it is made, reorganized and transformed (Swyngedouw, 1997).

Embracing the position that scale is made by and through social processes, within the context of particular histories and geographies, is a radical reconceptualization. The fourth edition of *The Dictionary of Human Geography* describes scale as:

> a central organizing principle according to which geographical differentiation takes place. It is a metric of spatial differentiation; it arbitrates and organizes the kinds of spatial differentiation that frame the landscape. As such it is the production of geographical scale rather than scale *per se* that is the appropriate research focus. (Smith, N., 2000: 725)

Most social theorists within geography use Henri Lefebvre's conceptualizations about the production of space as a prerequisite for understanding the social production of scale (Lefebvre, 1991).[1] Lefebvre's observation that space is a social product has encouraged geographers to specify the historically and geographically contingent social processes by which scale constructions contribute to the production of space. Arguably, the literature on the social production of scale can be divided into two camps: those who attempt the task of abstract theory building through close readings of Lefebvre or Marx, and those who have explored case studies in order to delineate moments of scale production. Both case studies and abstract theorizing have been valuable in their own ways, and both suggest that there are least three central tenets that currently constitute our understanding of scale production. First, scale is not a pre-existing category waiting to be applied, but a way of framing conceptions of reality. This means that different scales constitute and are themselves constituted through an historical-geographical structure of social interactions (Delaney and Leitner, 1997; Smith, N., 1992). Second, the particular ways in which scale is produced have material consequences. Scale making is not only a discursive practice, it is also the tangible outcome of the practices of everyday life as they articulate with and transform macro-level social structures. Third, scale productions are often contradictory and contested, and can easily be dismantled as new contextual factors become salient. Scale production is a political process, shaped as well by social, cultural and economic forces, and endemic to capitalism, making any particular scale construction always potentially open to further transformation. The quote by Erik Swyngedouw that opens this chapter elegantly summarizes these three tenets.

In attempting to situate the moments of scale production, scholars have mostly focused on how large scale structural transformations provide a necessary context for new scale productions. Contemporary globalization is the most popular historical context for exploring scale construction (Brenner, 1997a; Jessop, 1997a; Jessop, 1997b; MacLeod and Goodwin, 1999a: MacLeod and Goodwin, 1999b; Katz, Newstead, and Sparke, forthcoming). Not surprisingly, scholars have also argued that "geography matters" to scale production as unique culture, social practices, and political-economic forces at play in a place shape opportunities for and create barriers against it. Finally, there also appears to be substantial agreement in the literature – by way of the primary objects of analysis of the published work – that the key structural components in a theory of scale are the state and capital (Leitner, 1997; Miller, 1997; Brenner, 1997b; Brenner, 1998; Cox, 1998; Smith, N., 1984: Smith, N., 1995; Smith and Dennis, 1987; Taylor, 1982; Taylor, 1984: Taylor, 1987), nonstate level political actors such as labor (Herod, 1995; Herod, 1996; Herod, 1997), political parties (Agnew, 1993; Agnew, 1995; Agnew, 1997), political activists (Miller, 2000; Adams, 1996; Brown, 1996), and ensembles of urban actors known as "urban regimes" (Jessop, 1997a; Jessop, 1997b; MacLeod and Goodwin, 1999a; MacLeod and Goodwin, 1999b). Generalizing broadly, the research to date confirms that the constantly transforming scales that have become stabilized during this particular moment in capitalist history and geography are the outcome of a particular logic of capitalist expansion.

This focus on the social relations of production is hardly surprising given that the reorganization of the global economy of the last twenty-five years has relocated industrial production activities from core to periphery and enabled the emergence and predominance of service-based production in the core. Yet the reorganization of production is not the only profoundly central structural transformation of the last twenty-five years. Social reproduction – and particularly its decreasing salience in state practices – has also played a significant role in contemporary scale production. As a result of this preoccupation with questions of capitalist production in critical human geography, we have yet to untangle the full complexity of the social production of scale. I attempt to redress this oversight through a discussion of nineteenth-century scale production articulated through white, middle-class, urban, US women's movements addressing issues of social reproduction. I chose this historical geography because it is a period and place of extensive state institutional and capacity building around public service and welfare provision. It is also, relatedly, a period and place of previously unprecedented consumption opportunities and practices, as culture and economy cohered around an expanding US market for goods and services. An emerging, increasingly urbanized (and suburbanized) white middle class oriented its identity construction in no small part around household acquisition (Walker, 1981). Paralleling in many ways the turn of the twentieth century, the turn of the nineteenth century was a time of exceptionally consequential political, economic, social, and cultural restructuring in response to an emergent corporate industrial capitalism, fuelled by extensive foreign immigration within the context of accelerating and largely unregulated urbanization. An increasingly culturally homogenous urban middle class (located across the solidifying US Manufacturing Belt) was central to this historical moment particularly, with respect to the demands it made upon the state to mediate the negative impacts of the three interrelated processes of urbanization, industrialization, and immigration (Gordon 1978; Hays, 1974). These women's movements, organized broadly around a "progressive" agenda and articulating a consistent discourse, helped philosophically and physically to reorient the state to expand its purview and assume increasing responsibility for public service and social welfare provision (Schultz, 1989; Rosenkrantz, 1972).

The transformation of the US state at the turn of the nineteenth century signaled the appearance of a profoundly different popular understanding of its role at the local, state, and federal levels. The federal state's role was no longer confined to promoting commercial activity through subsidies and taxes.[2] By assuming responsibility for the social welfare of poor and elderly citizens, the federal state was reconfiguring its relationship to both capital and labor, and assuming new moral, political, and financial burdens that would require both new institutions and the creation of and ability to respond to new constituencies (Skocpol, 1992; Skowronek, 1982; Weibe, 1967). At the local level, the state also began to assume and create new institutions and capacities. Between 1870 and 1920, local governments institutionalized responsibility for providing the physical infrastructure of capitalism including schools, roads, public parks, sanitation works, clean water delivery systems, street lights and police protection (Schultz and McShane, 1988). In effect, the local state assumed a new role as the deliverer of collective consumption goods, prefiguring the crises that necessitated widespread capitalist state restructuring in the United States beginning in the 1970s.[3] This period is particularly instructive to explore as it provides both a prelude and a dramatic contrast to contemporary state reterritorialization and scale production, illuminating how social reproduction figured prominently in the institutional and capacity building practices of new US state forms.

Theorizing Social Reproduction

Social reproduction, as a feature of capitalist social relations, involves two key components: the *forces* of production and the *relations* of production. The productive forces reflect the actual material foundations that enable the production of goods and services and include such elements as the level of technological development and the quality and availability of human and physical resources. The relations of production incorporate the social structure through which these goods and services are produced. In capitalism, one group – the bourgeoisie – owns the productive forces (e.g., the machines, the telecommunication system, the finances) while the other, laborers, only control their labor power or their ability (e.g., physical strength, formal educational preparation, acquired skill). The social structure that configures the relations of production also includes the ways in which gender, race, age, religion, and other related variables shape social attitudes and behaviors.

Social reproduction involves maintaining labor power as a force *and* a relation of production. Most importantly, for my argument, it involves not only the production of goods and services but also the material conditions of daily existence that enable that production to occur. Cindi Katz writes that social reproduction entails "... the fleshy, messy, and indeterminant stuff of everyday life" (Katz, 2001). In short, social reproduction is a complex set of knowledges and practices that, while bound up with capitalist production, also reflect the bodily particularities of the times, places, and people within which and by whom they are enacted.

Feminist theorists have provided us with our most important insight about social reproduction: that the relationships, objects, and practices enabling the taken-for-grantedness of our everyday lives and constituting social reproduction are not outside the realm of theoretical relevance but are absolutely central to it (Benhabib and Cornell, 1987; Nicholson, 1987; Hartsock, 1984). As Dorothy Smith has made clear, the commonsense knowledge and seemingly unexceptional ways of doing and being in our daily lives – our social, physical, cultural, and emotional routines, customs, habits, and activities – are also at the theoretical and practical center of social reproduction (Smith, D., 1987). Thus, the microlevel infrastructure of the household, where labor power is refreshed and reproduced on a daily basis, is as inextricably bound up with the operations of capitalism as are capital and the state. The state is also certainly as much bound up with the operations of the household as it is with the operations of the economy. We must be careful, however, to recognize that the features of social reproduction, etched into the social relations of the household, are not purely or exclusively mediated by capitalism. While capitalism surely penetrates even the most intimate social practices, including sexual practices as queer theorists have shown (Butler, 1993; Joseph, 2002), other structural systems, especially cultural systems, also configure the every day world of the home.[4]

The gender relations that frame household relationships, interactions, and expectations, and that are constituted heterosexually through the cultural institutions of marriage and the family, are certainly among the most salient mediators of social reproduction in the home.[5] In the (heterosexual) home, the social relations of capitalism and of patriarchy are cohabiting, directly and indirectly shaping the practices of everyday life in intimate and obvious ways. Like other cultural systems, patriarchy – a mode of power relations that results in the subordination of women – is not expressed uniformly across time and space. This linking of capitalism and patriarchy in the nineteenth-century US home is central to my attempt to explicate the complexity of the social production of scale in any time or place. It enables me to explore two of the key theoretical and political components of late-nineteenth-century urban US social life, and the way they came together to shape social movements that forced a reconfiguration and elaboration of state functions within the context of a rapidly transforming system of production. Before I can proceed to that discussion, however, there is a component of social reproduction that needs further elaboration.

Marx argued that the consumption of commodities produced outside the household was a central part of capitalist production. Through consumption, people maintain themselves both physically and as members of a particular social class within a particular society. As a result, consumption occurs along both *physiological* and *historical* axes. The basic needs that must be met to sustain human existence – food, shelter, clothing – satisfy our physiological needs. Historically determined needs, however, emerge out of new developments in the forces of production and changing relations between social classes that are articulated through new cultural norms. "[W]ith the growth of large-scale industry, generalized mechanization of labor, constant differentiation of commodities, and growing physiological and nervous wear and tear of labor power, consumer goods become more and more determined by technical innovations and changes in the sphere of production" (Bottomore, 1983: 93). It is these historically determined needs, arising out of dramatic changes in US industrial capacity in the late nineteenth and early twentieth century, that are particularly important to my argument.

Social historians and feminist theorists note that the nineteenth-century US urban middle class used consumption as a way of constituting its identity. As Victoria de Grazia (1996, 18) contends: "the making of nineteenth-century class society was not only about transformations in the relations of people to the means of production but also about their massively changing relations to systems of commodity exchange and styles of consumption." This bourgeois mode of consumption became predicated on women as heads of households, who transformed households from purchasing largely to satisfy physiological needs, to purchasing as a "consuming class" and thereby constituting "its identity through a shared pattern of acquiring goods and a common structure of taste" (de Grazia, 1996: 18). As a result, "goods came to represent and even constitute people, groups and institutions in a new way. That is, class, gender, nation, and even self were constructed through the acquisition and use of goods" (Auslander, 1996: 81). The late-nineteenth/early-twentieth-century white, urban middle-class home was thus a complex site of dramatic new changes in production, social reproduction, and consumption. These structural transformations also effected broad spatial change outside the home, in the streets and neighborhoods of the burgeoning cities of the US manufacturing belt. In response to these changes, and operating through new discourses of class, race, and gender identity, significant numbers of white, urban, middle-class women became politically active outside the home, educating and exhorting local, state, and eventually federal governments to respond to the challenges and crises that an emerging corporate capitalist production regime was generating. Known broadly as "the progressives," this group, as well as substantial collectivities of white, urban, middle-class men, was remarkably successful in reshaping the functions and territorial responsibilities of the state to include the provision of public goods and social welfare. By the time the twentieth century was well under way, the US state had assumed a formal scalar configuration organized around its role as the manager of the urbanization of capital and the negative impacts of urban industrial production. The turn of the nineteenth century therefore, is a moment of a *fleshing out* of the state, rather than its more recent hollowing out, as well as ultimately the intensification of the political *internationalization* of the state as the United States reluctantly entered the arena of the World War I.

Domesticating the State

Urban historians have demonstrated that the turn of the nineteenth century was a pivotal moment in the urbanization of capital in the United States. As small scale firms, many of them rural in origin, were bought up and consolidated into larger and larger corporations, capital investment flowed increasingly into the twinned processes of industrialization and urbanization, drawing more and more native rural laborers and foreign immigrants to America's burgeoning cities. By the 1920 census, fewer people lived in rural areas than in urban areas and the United States was well entrenched on its path to become a thoroughly urbanized economy. Moreover, as David Harvey (1989) has pointed out, the urbanization of capital was accompanied by the urbanization of consciousness as new class fractions emerged as foreign immigrants arrived and people from rural backgrounds joined the urban/industrial workforce. This process created an unprecedented confrontation of social groups and ways of life in a largely unregulated context.

The political pressures upon the state to assume increasing responsibility for public and social welfare provision derived largely from increasingly vocal and well-organized women who, lacking the franchise, used other indirect means to influence the state. In the women's movement and its related strands – including voluntary motherhood, domestic feminism, municipal housekeeping, mother's aid and other social provision movements – US women, largely middle class and of western European decent, utilized the home as a site of political inspiration and engagement with some of the structural opportunities that the transforming political economy presented to them.

These women operated within a particular cultural context that widely recognized naturally ordained separate spheres of life for men and women. As Jürgen Habermas and his critics have argued, the public sphere of politics and the market belonged to men. Money and power were the coin of that realm. The private sphere of the home was understood to be the purview of women. Intimacy and affection were its coin (Habermas, 1989; Calhoun, 1992). Habermas's theorizing about the public sphere was based on eighteenth-century France. By the late nineteenth century in the United States, these separate spheres, though still operational were considerably transformed. Although in the traditional understanding of the public/private split, access to the economy and the state is available only to political citizens, legal persons, and economic agents, nineteenth-century middle-class women were effectively none of these (Marston, 1990). Yet women used their identities as mothers and housewives to assert a moral superiority and expert knowledge directed at resolving crises that were emerging around the urbanization of capital. In the following sections I outline two different movements that middle class urban women joined in large numbers at or around the turn of the nineteenth century. I use these movements to illustrate the existence of a discourse of maternalism and domesticity that empowered women to directly address the state in unprecedented ways. The first is the women's club movement. I focus on the particular case of women's clubs in Chicago, because the case provides a great deal of empirical detail on the interaction between social agents and the local state as it was transformed to address new and different responsibilities for the social reproduction of capital and labor. My second case is the federal state and the movement for widows' and soldiers' pensions, again because it provides a detailed illustration of the interaction of agents and structures in the production of new scales of state functions.

Other movements were also important in producing different scales of activism around maternalism and domesticity. These include the voluntary motherhood and social purity movement, aimed at garnering more control for women over fertility, sexuality, and childrearing (Gordon, 1976; Smith-Rosenberg, 1982). Also significant was the domestic feminism movement that I describe at length elsewhere (Marston, 2000). This movement was directed at providing women a greater degree of authority over the household and the social relations of childrearing. Space limitations prevent me from discussing either of these movements in detail here.

It is important to understand that the during the late nineteenth century, the US middle class was attempting to consolidate its position as a dominant economic, social, political, and cultural force. Increasing in numbers due to the opportunities that industrialization and urbanization opened up for new "professional" occupations, and responding to the perceived threat that new immigrants posed to their social and cultural identities and their economic power, the US middle class began to enact its demands through direct and indirect political means. It is also important to point out that the constitution of the middle class as an important social form was articulated by white, urban, middle-class women in the United States through the spatial form of the middle-class household, where very particular ideas about everything from child-rearing to table-setting to sexual intercourse were being formulated, practiced, and deployed through a range of vehicles from women's magazines to "Americanization programs" to school curricula. In effect, the middle-class progressives were calling upon the state to discipline the city and its new residents to adhere to their material and discursive practices as an increasingly dominant class. In short, the middle class became the dominant social form of the period; and the middle-class home became the dominant spatial form.

The club movement and municipal housekeeping

At the same time that Victorian women were advancing fundamental changes in attitudes and behaviors about the female body and the home, they were also revising attitudes about the permissibility of women's participation in activities outside the home. Interestingly, arguments about the moral superiority of women and the cult of motherhood that surrounded the voluntary motherhood and social purity movements were rehearsed in the women's club movement (Baker, 1984; Blair, 1984; Hayden, 1981). Blossoming in the United States from the late 1870s and 1880s, the club movement enabled women "to leave the confines of home without abandoning domestic values" (Blair, 1980: 5). Initially, the clubs were devoted to cultivating and promoting opportunities for women's self-development, mostly through literary and current events discussions. Eventually, the clubs helped to cultivate new ideologies about women's moral imperatives, and women began to invoke their "natural" talents as a justification for entering and attempting to alter the market-based, competitive public (male) sphere. In doing so, white, urban, middle-class women emphasized that their natural abilities for domestic affairs and their increasing technological sophistication could be put to good use in a society whose "public house" was disordered, unsanitary, and lacking in adequate cultural and recreational opportunities.

Maureen Flanagan has written incisively on the progressive era club movement, demonstrating how gendered identities shaped the kinds of reforms that the women urged upon the city of Chicago (Flanagan, 1990). Using membership and activities records from both The City Club (the men's civic reform organization) and the Chicago Woman's City Club (the female counterpart), Flanagan reveals the discursive framework around which these women came to see the city as a logical extension of the home, and therefore saw themselves as "well-placed" to offer expert advice on how to address its problems. The statement of purpose of the Woman's City Club of Chicago taken from a 1911 *Bulletin* reads:

> To bring together women interested in promoting the welfare of the city; to coordinate and render more effective the scattered social and civic activities in which they are engaged; to extend a knowledge of public affairs; to aid in improving civic conditions and to assist in arousing an increased sense of social responsibility for the safeguarding of the home, the maintenance of good government, and the ennobling of the larger home of all – the city.

Despite their disenfranchisement, Flanagan's study shows that the club-women were actively involved as political reformers, making demands on the government of the city of Chicago that thoroughly reoriented its traditional purview. The Woman's City Club members were persistently and effectively involved in shaping the way that private power was wielded for public purposes.

Flanagan demonstrates that men's and women's clubs approached the solution to the city's myriad problems in very different ways, despite coming from the same solidly middle to upper-middle class and the same neighborhoods. Indeed they were often directly related to each other by marriage or birth. Yet, the clubs took opposing positions on all of the municipal issues that Flanagan investigated. While the City Club consistently offered private solutions to public problems that would protect and further the aims of business, the Woman's City Club of Chicago offered proposals that would solve public problems through public intervention, directed at the well-being of everyone in the city regardless whether they were a business or an orphaned child.

These different positions were reflected by the different experiences of each group due to their distinct daily lives. The Clubmen were businessmen, accustomed to thinking what was best for business was best for the city. The Clubwomen were used to operating in the home and ensuring the well-being of all family members.[6] This same objective informed their city-wide agenda. Flanagan argues that the very term that came to characterize women's participation in the progressive reform movement, "municipal housekeeping," captures the intimate connections between home and city as well as between citizens and government. The term also enabled women to bridge the ideological and practical gap between public and private life by pointing out that urban affairs were part of the home. Flanagan (1990: 1050) says it best: "by depicting the city as the larger home, the women were asserting their right to involve themselves in every decision made by the Chicago city government, even to restructure that government."

Flanagan's excellent work illustrates the importance that the Woman's City Club of Chicago – and other groups like it in the US Manufacturing Belt – played in structuring the local state around expanded and new concerns for what Manuel Castells (1977) would call "collective consumption," or others would call public goods. Men and women progressives had different and significant impacts on local governments through a range of reforms, from the application of business management principles to government to the municipal provision of services for the health and welfare of all citizens. They were part of a turn of the century *urban regime* that reoriented the local state in substantial ways, enlarging its functions and responsibilities and creating a new scale of state territoriality in the process. It was women who were particularly effective, however, in constructing a local state scale around concerns of social reproduction.

The federal state and social welfare provision

In the early decades of the twentieth century, women were also working steadily and successfully at transforming the federal state to assume new responsibilities for social welfare provision. There are several excellent texts on this period and the role of women in pushing for the construction of a "welfare state." I draw particularly on the work of Theda Skocpol (1992), Molly Ladd-Taylor (1994), Linda Gordon (1994), and Gwendolyn Mink (1995). They provide four careful and enlightening treatments of the ways that early-twentieth-century federal state functions and institutions were altered and augmented in response to pressure from women's groups, and influential women, who argued that it was the government's responsibility to ensure the social well-being of its citizens, especially women and children.

Between 1910–20, women's groups were able to push state and local governments to establish programs to aid single mothers, effectively creating the first program of modern public welfare. Mother's aid became so well-accepted, as a concept and a practice, that all but two US states passed laws creating it. The federal government nationalized it by the end of the

Depression, creating support for women with dependent children based on the mother's aid model. It is widely accepted among scholars of gender and the modern US state that the contemporary welfare program of Aid to Dependent Children (ADC) originated in 1915 with these state-level mother's pension programs.

The creation and implementation of social welfare policies by all levels of the state emerged out of the same discourse of maternalism that informed the other women's movements already discussed. As Mink and others have shown, the mother's aid policy activists drew "from the republican ideology of gendered citizenships to the separate spheres ideology of the Victorian period to the mothers' politics of the late nineteenth and early twentieth centuries which asserted women's political significance to "the race" and the nation" (Mink, 1995: 32). By linking the private with the public, social and policy activists connected the health of the polity to the quality of motherhood, and demanded that government provide economic assistance to poor women and children. The primary aim of the maternalist social activists was child welfare and they pointed to home conditions as well as educational opportunities as its measures. Ladd-Taylor summarizes maternalism as an ideology and a discourse whose adherents hold: "(i) that there is a uniquely feminine value system based on care and nurturance; (ii) that mothers perform a service to the state by raising citizen-workers; (iii) that women are united across class, race, and nation by their common capacity for motherhood and therefore share a responsibility for all the world's children; and (iv) that ideally men should earn a family wage to support their "dependent" wives and children at home" (Ladd-Taylor, 1994: 3). It should be recalled that industrialization, immigration, and the movement of freed slaves into northern and midwestern cities, together were creating cultural anxieties among "old stock" Americans.[7] Late-nineteenth-century scientific racism contended that excessive cultural diversity corrupted citizenship and threatened the stability of democracy. Whereas some Americans mobilized to restrict immigration, and segregate those who were already here, maternalist reformers operated according to an assimilationist perspective and expected to incorporate the new groups into white, middle-class norms, institutionalized through public policies and programs directed at poor (new immigrant and black) women and children. Mother's aid policies were a way of reforming the socialization of poor urban women and children and, as such, acted as an effective form of social control.

By the end of the 1920s, mother's aid advocates and supporters had dramatically altered the political landscape. While mother's aid moved the state more toward charitable work than outright welfare, it did condition both the citizenry and politicians to the notion of public welfare. As Linda Gordon argues:

> Mother's aid politics also produced at least two other major legacies for the New Deal and contemporary welfare policy. Along with a variety of other Progressive Era reform causes, it accelerated the development of interactive national networks of women welfare reformers and professional social workers. It promoted social work as a key route to political influence for women, and set guidelines for welfare state development.... The whole experience of mother's aid, from its conception to the evaluation of its administration, helped congeal a particular view of public welfare which conditioned, more than any other single factor, the future shape of ADC. (Gordon, 1994: 64)

The mother's aid movement was undoubtedly an important social movement as it was responsible for the establishment of the US Child Bureau, the US Women's Bureau, and eventually, through New Deal legislation, the foundations for modern welfare directed at women and children and the extension of pensions and benefits to veterans and the elderly. These were all key components in the development of the US welfare state.

The two movements described above, as well as the voluntary motherhood/social purity and domestic feminism movements, enabled turn of the century middle-class, white, US urban women to configure a scalar fix premised on maternalism and domesticity. Operating through this discursive framework, these women very successfully reoriented the functions and purview of the state so that it came increasingly to assume responsibility for, and construct legislation aimed at, preserving and enhancing the social reproduction of its citizenry around the norms of the white, urban, middle class. Certain scales were central to this reorganization of the local, state, and federal state: the body, the home, and city, and the nation.

As the twentieth century unfolded, it is possible to trace a decline in relevance of the maternalist/domesticity discourse and of its influence over state functions and organization. Major events such as the Great Depression, the two World Wars, and the Cold War, as well as prominent social movements such as the environmental movement, the consumer movement, and the 1960s and 1970s women's movement inscribed revised scales of state organization and function.

Conclusion

One of the most striking aspects of the US women's reform movements at the turn of the nineteenth century is the way that maternalism and the home provided ideological and material bases from which white, middle-class, urban women were able to launch an effective and far-reaching political agenda. The home helped women to connect ideas and practices across realms that might seem discrete on the surface and are often kept separate by the larger structuring forces of society such as capitalism and patriarchy. Operating through a changing discourse of motherhood and "homemaking," Victorian and early-twentieth-century white, urban, middle-class women in the United States pushed for reforms at all levels of government aimed at redressing widespread and acute problems of social reproduction. As a result, the federal state was moved from a narrow focus on facilitating trade and commerce to one that embraced the support and maintenance of its most disenfranchised citizens. Prefiguring the reorientation of the federal state to a social welfare framework was the reorganization of the functions and institutions of government at local and state levels. The local state assumed enlarged responsibilities for providing the physical infrastructure for capitalist expansion, and extended new services and programs to its citizens in bundles of publicly delivered common goods. The US state government also became involved in social welfare provision, furnishing a foundation for the later implementation of the Federal government's new programs for women and children.

Narrating the history of women's turn of the century activism around social reproduction has allowed me to explore scale production – political and economic territorial restructuring – as an outcome of the tensions between structures and agents in the performance of routine as well as extraordinary practices. Certainly by the mid-1970s, the aims of the turn of the nineteenth century's women's movements described here had begun to seem irrelevant, insupportable, or normalized into the state's activities. And, as the recent research on scale production has amply shown, by the turn of the twentieth century, a radically new scalar fix was being negotiated and produced. Yet this empirical focus on the turn of the nineteenth century provides a helpful counterpoint to the extensive discussions of globalization that dominate contemporary analyses of the production of scale. At the start of the twenty-first century, social reproduction in the United States has become dramatically privatized as the state continues to abandons its century-old role as provider of the public goods and services that enable the social reproduction of labor and production. The turn of the nineteenth century in the urban-industrial United States was a time of mounting social pressure directed at the state to enjoin it to assume increasing responsibility for social reproduction; pressures that eventually spilled over into the collective consumption crises of the 1970s. Whereas most studies of the social production of scale start with the 1970s world economic crisis, I would argue that it is particularly valuable to begin at least a century earlier.

Notes

1 Brenner also relies quite heavily for his theorization of scale on Lefebvre's four volume *De L'État*.

2 Until the New Deal, a unique system of federalism determined the role of the national government in American life. State governments were responsible for deciding the contours of the rights, privileges, and obligations of their citizenry through state laws and policies that organized and regulated daily life including issues of education, property, family, commerce and labor, banking and criminality. The national government, through the Constitution, was restricted from interfering in such activities as its role and powers were limited to the regulation of commercial activities between states. In the late nineteenth and early twentieth century, the role of the federal government began to be challenged as citizens agitated for more centralized administrative functions. Ultimately, in response to the Great Depression and the demands of social movements, the federal government, under Franklin Roosevelt and his New Deal policymakers, enacted a constitutional revolution by creating regulatory labor policies and redistributive social policies that led to an expanded federal government presence in the lives of all Americans. The New Deal and its impact on federal state functions was the culmination of years of social pressures initiated during the Progressive Era that directed government responsibility for social citizenship first to the municipal and later to state level.

3 Manuel Castells (1972) is responsible for this conceptualization of the city as a space of collective consumption. But see also Cynthia Cockburn (1977), Mark Gottdiener (1987), and James O'Connor (1973) for treatments of the local state in crisis in the 1970s over the provision of collective consumption goods.

4 Of course, it is not just the home that is shaped by culture. The factory floor, the boardroom, the playground, and the marketplace are all structured by particular cultural practices and meaning systems as well as by the political economy of capitalism.

5 This is not to dismiss the importance of gender relations in the operation of workplace social relations or to ignore age or sexuality as other possible mediating frameworks in the home. I am simply highlighting those conceptual aspects of social reproduction that are critical to the central thread of my argument.

6 Flanagan also found that the two clubs approached the gathering of data about the city's problems very differently. The men, in their corporate positions of management and authority, were used to experiencing only parts of business problems first hand as employees often gathered information for them. They used the same techniques to gather data about the city's ills. The women, used to experiencing and solving household problems directly, organized their members to go out and investigate problems by observing them and talking to the people who were experiencing them.

7 Old stock Americans are considered those who settled during the first wave of European immigration to the United States including Americans of English,

Welsh, Scottish, Irish, and German extraction. New immigrants came from Italy and parts of Eastern Europe and were mostly Jewish or Catholic.

References

Adams, P. C. 1996: Protest and the scale politics of telecommunications. *Political Geography*, 15: 419–42.

Agnew, J. 1993: Representing space: Space, scale and culture in social science. In J. Duncan and D. Ley (eds), *Place/Culture/Representation*, London: Routledge, 251–27.

Agnew, J. 1995: The rhetoric of regionalism: The Northern League in Italian Politics, 1983–1994. *Transactions of the Institute of British Geographers*, n.s.20: 156–172.

Agnew, J. 1997: The dramaturgy of horizons: Geographical scale in the "reconstruction of Italy"by the new Italian political parties, 1992–95. *Political Geography*, 16: 99–122.

Auslander, L. 1996: The gendering of consumer practices in nineteenth-century France. In V. de Grazia (ed.), *The Sex Of Things: Gender And Consumption In Historical Perspective*, Berkeley, CA: University of California Press, 79–112.

Baker, P. 1984: The domestication of politics: Women and American political society, 1780–1920. *American History Review*, 89: 620–47.

Benhabib, S. and Cornell, D. 1987: Introduction: Beyond the politics of gender. In S. Benhabib and A. Cornell (eds), *Feminism As Critique*, Minneapolis, MN: University of Minnesota Press, 1–15.

Blair, K. 1980: *The Clubwoman as Feminist: True Womanhood Redefined, 1868–1914*. New York: Holmes & Meier.

Bottomore, T. (ed.) 1983: *A Dictionary of Marxist Thought*. Cambridge: Harvard University Press.

Brenner, N. 1997a: Global, fragmented, hierarchical: Henri Lefebvre's geographies of globalization. *Public Culture*, 10: 135–67.

Brenner, N. 1997b: State territorial restructuring and the production of spatial scale: Urban and regional planning in the Federal Republic of Germany, 1960–1990. *Political Geography*, 16: 273–306.

Brenner, N. 1998: Between fixity and motion: Accumulation, territorial organization and the historical geography of spatial scales. *Environment and Planning D: Society and Space*, 16: 459–81.

Brown, M. 1996: Sex, scale and the "new urban politics": HIV-prevention strategies from Yaletown, Vancouver. In D. Bell and G. Valentine (eds), *Mapping Desire: Geographies Of Sexualities*, London: Routledge, 245–63.

Butler, J. 1993: *Bodies that Matter: On the Discursive Limits of "Sex."* London & New York: Routledge.

Calhoun, C. (ed.) 1992: *Habermas and the Public Sphere*. Cambridge, MA: MIT Press.

Castells, M. 1977: *The Urban Question: A Marxist Approach.* Cambridge, MA: MIT Press.

Cockburn, C. 1977: *The Local State.* London: Pluto Press.

Cox, K. 1998: Representations and power in the politics of scale. *Political Geography,* 17, 41–4.

de Grazia, V. 1996: Introduction. In V. de Grazia (ed.) *The Sex of Things: Gender and Consumption in Historical Perspective,* Berkeley, CA: University of California Press, 11–24.

Delaney, D. and Leitner, H. 1997: The political construction of scale. *Political Geography,* 16: 93–7.

Flanagan, M. 1990: Gender and urban political reform: The City Club and the Woman's City Club of Chicago's progressive era. *American History Review,* 95: 1032–50.

Gordon, L. 1976: *Women's Bodies, Women's Rights: Birth Control in America.* New York: Penguin Books.

Gordon, L. 1994: *Pitied But Not Entitled: Single Mothers and the History of Welfare, 1880–1935.* New York: The Free Press.

Gordon, D. 1978: Capitalist development and the history of American cities. In W. K. Tabb and L. Sawers (eds), *Marxism and the Metropolis,* New York: Oxford University Press, 25–63.

Gottdiener, M. 1978: *The Decline of Urban Politics.* Newbury Park, NJ: Sage.

Habermas, J. 1989: *The Structural Transformation of The Public Sphere.* Translated by Peter Burger & F. Lawrence. Cambridge: MIT Press.

Hayden, D. 1981: *The Grand Domestic Revolution: A History of Feminist Designs for American Homes, Neighborhoods, and Cities.* Cambridge, MA: MIT Press.

Hartsock, N. 1984: *Money, Sex and Power.* Boston, MA: Northeastern University Press.

Harvey, D. 1989: *The Urban Experience.* Baltimore, MA: Johns Hopkins University Press.

Hays, S. P. 1974: The changing political structure of the city in industrial America. *Journal of Urban History,* 1: 6–38.

Herod, A. 1995: International labor solidarity and the geography of the global economy. *Economic Geography,* 71: 341–63.

Herod, A. 1996: Labor as an agent of globalization and as a global agent. In K. Cox (ed.), *Spaces of Globalization,* New York: Guilford, 167–200.

Herod, A. 1997: Labor's spatial praxis and the geography of contract bargaining in the US East Coast longshore industry, 1953–1989. *Political Geography,* 16: 145–70.

Holly, B. 1978: The problem of scale in time-space research. In T. Carlstein, D. Parkes, and N. Thrift (eds), *Timing Space And Spacing Time,* vol. 3. New York: Wiley, 108–38.

Hunt, C. L. 1980 (1912): *The Life of Ellen H. Richards.* Washington, DC: American Home Economics Association.

Jessop, R. 1997a: The entrepreneurial city: Re-imaging localities, re-designing economic governance, or restructuring capital? In N. Jewson and S. MacGregor (eds), *Transforming Cities: Contested Governance and New Spatial Divisions*, London: Routledge, 28–41.

Jessop, R. 1997b: A neo-gramscian approach to the regulation of urban regimes: Accumulation strategies, hegemonic projects, and governance. In M. Lauria, (ed.), *Reconstructing Urban Regime Theory*, London: Sage, 51–73.

Joseph, M. 2002: *Against the Romance of Community*. Minneapolis, MN: University of Minnesota Press.

Katz, C., Newstead, C., and M. Sparke 2003: Ideology, resistance and the cultural geography of scale-jumping. In S. Pile (ed.), *The Cultural Geography Reader*. London: Sage.

Katz, C. 2001: Vagabond capitalism and the necessity for social reproduction. *Antipode*, 33: 709–28.

Kirby, A. 1985. Pseudo-random thoughts on space, scale, and ideology in political geography. *Political Geography Quarterly*, 4: 5–18.

Ladd-Taylor, M. 1994: *Mother-Work: Women, Child Welfare, and the State, 1890–1930*. Urbana & Chicago, IL: University of Illinois Press.

Lefebvre, H. 1976: *De L'État*. 4 vols. Paris: Union Générale d'Éditions.

Lefebvre, H. 1991: *The Production of Space*. Oxford: Basil Blackwell. Translated by Donald Nicholson-Smith.

Leitner, H. 1997: Reconfiguring the spatiality of power: The construction of a supranational migration framework for the European Union. *Political Geography*, 16: 123–44.

MacLeod, G. and Goodwin, M. 1999a: Space, scale and state strategy: Rethinking urban and regional governance. *Progress in Human Geography*, 23: 503–27.

MacLeod, G. and Goodwin, M. 1999b: reconstructing an urban and regional political economy: On state, politics, scale, and explanation. *Political Geography*, 18: 697–730.

Marston, S. 1990: Who are "the people"?: Gender, citizenship, and the making of the American nation. *Environment and Planning D: Society and Space*, 8: 449–58.

Marston, S. 2000: The social construction of scale. *Progress in Human Geography*, 24: 219–42.

Marston, S. and Smith, N. 2001: States, scales, and households: Limits to scale thinking? *Progress in Human Geography*, 25: 615–19.

Miller, B. 1997: Political action and the geography of defense investment: Geographical scale and the representation of the Massachusetts miracle. *Political Geography*, 16: 171–85.

Miller, B. 2000: *Geography and Social Movements*. Minneapolis, MN: University of Minnesota Press.

Mink, G. 1995: *The Wages of Motherhood: Inequality in the Welfare State, 1917–1942*. Ithaca, NJ: Cornell University Press.

Nicholson, L. 1987: Feminism And Marx: Integrating kinship with the economic. In S. Benhabib and D. Cornell (eds.), *Feminism as Critique*, Minneapolis, MN: University of Minnesota Press, 16–30.

O'Connor, J. 1973: *The Fiscal Crisis of the State*. New York: St. Martin's Press.

Richards, E. 1881: *The Chemistry of Cooking and Cleaning: A Manual for Housekeepers*. Boston: Estes & Lauriat.

Richards, E. 1887: *Home Sanitation: A Manual for Housekeepers*. Boston: Ticknor & Co.

Richards, E. 1899: *The Cost of Living as Modified by Sanitary Science*. New York: John Wiley & Sons.

Richards, E. 1901: *The Cost of Food: A Study in Dietaries*. New York: John Wiley & Sons.

Richards, E. 1905: *The Cost of Shelter*. New York: John Wiley & Sons.

Rosenkrantz, B. 1972: *Public Health and the State: Changing Views In Massachusetts, 1842–1936*. Cambridge, MA: Harvard University Press.

Schultz, S. 1989: *Constructing Urban Culture: American Cities and City Planning, 1800–1920*. Philadelphia, PA: Temple University Press.

Schultz, S. and McShane, C. 1984: To Engineer the Metropolis: Sewers, Sanitation and City Planning in Late Nineteenth Century America. In R. Mohl (ed.), *The Making of Urban America*, Wilmington: Scholarly Resources, 81–98.

Skocpol, T. 1992: *Protecting Mothers and Soldiers: The Political Origins of Social Policy in the United States*. Cambridge, MA: Harvard University Press.

Skowronek, S. 1982: *Building a New American State: The Expansion of National Administrative Capacities, 1877–1920*. New York: Cambridge University Press.

Smith, D. 1987: *The Everyday World as Problematic: A Feminist Sociology*. Boston, MA: Northeastern University Press.

Smith, N. 1984: *Uneven Development: Nature, Capital and the Production of Space*. Oxford and Cambridge: Basil Blackwell.

Smith, N. 1992: Geography, difference and the politics of scale. In J. Doherty, E. Graham, and M. Mallek (eds) *Postmodernism and the Social Sciences*, London: Macmillan, 57–79.

Smith, N. 1995: Remaking scale: Competition and cooperation in prenational and postnational Europe. In H. Eskelinen and F. Snickars (eds), *Competitive European Peripheries*, Berlin: Springer, 59–74.

Smith, N. 2000: Scale. In R. Johnston, D. Gregory, G. Pratt and M. Watts (eds) *The Dictionary of Human Geography*, Oxford: Blackwell, 724–7.

Smith, N. and Dennis, W. 1987: The restructuring of geographical scale: Coalescence and fragmentation of the northern core region. *Economic Geography*, 63, 160–82.

Smith-Rosenberg, C. 1985: *Disorderly Conduct: Visions of Gender in Victorian America*. New York: Oxford University Press.

Swyngedouw, E. 1997: Excluding the other: The production of scale and scaled politics. In Roger Lee and Jane Wills (eds), *Geographies of Economies*, London: Arnold, 167–76.

Taylor, P. 1982: A materialist framework for political geography. *Transactions of the Institute of British Geographers*, n.s.7: 15–34.

Taylor, P. 1984: Introduction: Geographical scale and political geography. In P. Taylor and J. House (eds), *Political Geography: Recent Advances and Future Directions*, London: Croom Helm, 1–7.

Taylor, P. 1987: The paradox of geographical scale in Marx's politics. *Antipode*, 19: 287–306.
Walker, R. 1981: A theory of suburbanization: Capitalism and the construction of urban space in the United States. In M. Dear and A. J. Scott (eds), *Urbanization and Urban Planning in Capitalist Society*, London: Methuen, 383–429.
Weibe, R. H. 1967: *The Search for Order, 1877–1920*. New York: Cambridge University Press.
Yost, E. 1943: *American Women of Science*. New York: J.B. Lippincott.

9 Scale Bending and the Fate of the National

Neil Smith

In the last months of 1997, with financial capitalists across the world panicking about the wave of global economic crisis reverberating out of Thailand, Hong Kong, and Indonesia, several seemingly disparate but curious events turned up in the media. In the first story, New York's Mayor Rudy Giuliani, angry at the abandon with which UN diplomats seemed to flout local parking laws and blaming them for much of Manhattan's gridlock, threatened to begin towing illegally parked cars with diplomatic plates. Openly derided with the nickname Benito Giuliani for his erratic authoritarianism, the mayor raged even more angrily at the US State Department which, he felt, simply capitulated to this vehicular malfeasance. Maybe it has come to the point, Giuliani huffed, where New York City needs to have its own foreign policy. Four years later, in the wake of the September 11 tragedy, Giuliani found himself addressing the United Nations with a speech on New York's role in the world.

The United Nations also featured in the second set of headlines. The Atlanta media capitalist Ted Turner, erstwhile owner of CNN, announced in 1998 that in light of the UN's financial plight (caused in no small part by US refusal to pay its dues) he was donating them a billion dollars. Almost as generous was billionaire financier George Soros who responded to the imminent bankruptcy of the new capitalist state in Russia by providing $US 1/2 billion in loans to a desperate Yeltsin government. That figure was five times larger than the aid package offered by the US government that year. At about the same time the Disney Company (Turner's chief competitor) planned the release of a movie championing the religious monarchy of Tibet against Chinese military brutality, but they were apprehensive about possible Chinese government reaction. To smooth any ruffled feathers, Disney appointed Henry Kissinger as its "ambassador to China."

For Disney after all, a billion consumers is a terrible thing to waste (Katz, 1997).

One further event around the same time did not make such headlines. Based in the Philadelphia neighborhood of Kensington, the Kensington Welfare Rights Union (KWRU) had been organizing for years on a platform that insisted on welfare rights for poor people in the city. They argued powerfully that access to housing and especially a job represented a basic human right. They petitioned the Philadelphia City Council to no avail and so took their struggle to the state capital in Harrisburg. Rebuffed there, they went to Washington demanding that the Federal government provide decent-paying jobs, housing and other vital services for welfare recipients and poor people. Again rebuffed, they took their case to the United Nations. Hundreds of KWRU members and supporters set off from North Philadelphia to walk the hundred miles to UN Plaza on Manhattan's East Side. There they argued that the economic plight of poor people in the United States resulted from a class, race, and gender discrimination endemic to that society, and as such contravened the 1948 UN Declaration of Human Rights.[1] Downplayed in the mainstream media, this was nonetheless an embarrassing international indictment of the US government by some of its own citizens, on its own soil.

Something very odd is happening here. Any one of these events challenges our traditional sense of the proper role of city governments, nation-states, global corporations, and private individuals. Cities and states are not supposed to have their own foreign policy, presumably the prerogative of national states. Private individuals are not supposed to dwarf nation-states in bankrolling other national and transnational state institutions. In the home of the free, "domestic" activists are not supposed to jump scale and appeal to international authority for the resolution of local complaints. And since when did global corporations displace nation-states as the proper purveyors of diplomatic emissaries? Taken together, these events suggest intense "scale bending" in the contemporary political and social economy. Entrenched assumptions about what kinds of social activities fit properly at which scales are being systematically challenged and upset. This raises a host of questions. Why is this happening now? What are the causes? Is scale bending symptomatic of new geographies of capitalist expansion, and if so, what do they look like? How does scale bending affect particular scales inherited from past geographies? Who wins and who loses in the process? Finally, how does the restructuring of scale rework the landscape of empowerment and disempowerment for different classes, races, and genders of people?

These examples suggest not simply an economy gone global, the readjustment of local and national governments to so-called globalization, or even the emergence of a new class of cowboy capitalists scouting out the global frontiers of total capitalization. Woven through all of these shifts is a much more profound, multidimensional restructuring of the very geographical scales according to which the social economy is organized. These signs of scale bending are only the most obvious expressions of the reorganization of spatial difference that is currently underway in the global political and social economy. Restructured scales are a central metric of this spatial reorganization. But scale is of much more than momentary importance. As I have tried to argue previously, the question of scale is intrinsic to capitalism in a way unprecedented in previous modes of production. Socially construed scales of activity obviously preceded the capitalist mode of production, and will presumably outlive it, but as Marx demonstrated, the dialectic of expansion and centralization of capital becomes definitive of the capitalist mode of production. With capitalism, scale for the first time provides a vital geographical solution to this potential contradiction between expansion and centralization. The establishment of capitalism was from the start a construction of scales and scale differences, its uneven development is premised on the ability to construct and dismantle scales (and much more systematically so after the period from 1880 to 1919), and every restructuring of capital is a social and political restructuring of scale (Smith, 1990). The intense interest in geographical scale today, from issues of the body to those of the globe, is thus only understandable as a direct expression of the transformation of modern capitalism.

Naturalization, Gestalt, Lefebvre

There are many different conceptions of scale – representational, operational, descriptive, cartographic, and so forth – but until recently, the treatment of scale in liberal geographic and scientific research has tended to follow one of two specific paths. Scale is treated either as naturally given or as a methodological choice. In the first treatment, a certain hierarchy of scale is simply assumed. In the physical and biological sciences, for example, the span of scales from subatomic to atomic through molecular and cellular all the way to the scale of the universe is broadly treated as inherent to the structures of nature. Quarks, neutrinos, atoms, or solar systems simply come in the sizes they do, for all that they may expand or contract. Similarly in the social sciences, the scales of the body and the home, urban and regional, national, and global are also widely treated as given. In the second treatment, scale appears to be the obverse, namely a methodological choice: what is the appropriate scale of analysis for a specific piece of research? At which scale are data to be gathered? How is a reality at one scale to be represented at another scale, as in the drawing of a map?

These paths are certainly different but they are not mutually exclusive. In positivist science there is no necessary contradiction between voluntarist and naturalist assumptions about scale. Indeed, a voluntarism concerning scale – scale as choice – is necessarily balanced by gestures in the direction of an ontology of scale. Insofar as scientific investigation is assumed to leave the object of research unmodified, the voluntarism of the methodological approach actually abets the naturalization of scale.

Yet the mutual dependence of voluntarist and ontological treatments of scale is not foolproof. The choice of different scales of investigation can lead to very different kinds of statements about the realities being researched. One simple geographical illustration of this point concerns the scale of twentieth-century urbanism. It is conventional to argue that, with expansive suburbanization, twentieth-century urbanism experienced a dramatic decentralization from traditional urban cores. But this argument depends on an initial perspective that takes the urban center for the city itself. If instead we take view the city as a whole, for example from a circumplanetary satellite, the expansion of Los Angeles, Sâo Paulo or Shanghai in the twentieth-century landscape appears as an extraordinary *centralization* of social activity into existing urban centers, driven by economic expansion.

This "gestalt of scale" – the same object taken as a whole can look radically different from different scalar positions – haunts the conventional conceptual offset of voluntarist and ontological treatments of scale (Smith, 1987; Swyngedouw, 1997). Empirical associations that register statistically at one scale frequently look very different at different scales of analysis: by altering the chosen scale of vision a reality seen one way can suddenly appear as its opposite. Scale may be a choice, therefore, but different methodological choices of scale can generate radically different assertions about reality. Thus the gestalt of scale is not merely a procedural inconvenience (or convenience), but an endemic contradiction in liberal scientific approaches that insist on a definitively knowable reality.

In recent years the naturalization of scale has been challenged intellectually, much as scale bending has appeared in the political, social, cultural, and economic landscapes. Scale today is widely conceived in terms that hold the extremes of ontological and voluntarist treatments at bay for sake of arguably more complex approaches. Spurred by rapid technological innovations in the development, handling, and representation of geographical data, for example, many geographers have begun to treat scale as problematic, more malleable than in the past. Thus Dale Quattrochi and Michael Goodchild (1997) preface their review of scale in geographical mapping technologies with the ambition to develop a means for the "management and manipulation of scale." The mode is technocratic, aimed at improved geographic information for the improved management of public policy, and the positivist framework keeps intact the broad nexus of methodological and ontological assumptions about scale, but it also registers a heightened awareness of scale's plasticity.

As Sallie Marston (this volume) argues, however, it is social theorists who have been at the forefront of the recent, intense theoretical interest in geographical scale. For them, the "rejection of scale as an ontologically given category" is much more complete, giving way to a constructionist vision of geographical scale. She points back to Henri Lefebvre who, although not a theorist of scale, introduced the radical notion of the social production of space and provided a vital foundation for subsequent theorizations of scale (Marston, 2000: 220). The brilliance of Lefebvre's "production of space" argument lies in the fact that it contravenes the conceptions of space that have dominated western thought for two or three centuries, and opens the way for a thoroughly repoliticized conception of space. Challenging the notions of space embedded in western thought by Newton, Descartes, and Kant among others, he argues that space is not a given arena within which things happen, but the physical, social, and conceptual product of social and natural events and process. It is not that such events and processes take place "in space" but rather that space and spaces are produced as an expression of these social and natural processes. Social, mental, and physical space comprise a unity in this process. Lefebvre was not working in a vacuum, of course. A heightening of spatial language (of which Lefebvre was selectively critical) was already evident in French social theory, and English-language geography had also embarked on a critique of traditional western notions of space. But he was the most evocative, arguing that the production of space *is* the making of a political world, and vice-versa, and he came to decry the abstraction of space executed under the expansionist impress of capital accumulation and the capitalist state. Against this homogenization of space he championed "differential space," in which there is a direct translation from democratized social and political interests to democratized geographies, and his analytical optimism pointed toward a postcapitalist production of space. But Lefebvre was typically vague about the analytical roots and entailments of this differentiation of space. How does the differentiation of space – subversive or symptomatic of capitalist social relations – take place within a homogenizing capitalism? How would it work in a socialist world?

Attempting to answer these questions leads us in the direction of geographical scale. The production of geographical scale provides the organizing framework for the production of geographically differentiated spaces and the conceptual means by which sense can be made of spatial differentiation. The always malleable systems of geographical scales fix social differences temporariliy in more or less hierarchical spatial configurations. Or as Erik Swyngedouw (1992a) has argued, spatial scale represents a kind of territorial infrastructure, or geographical technology, for the expansion and reproduction of capital. This "technology" may be economically or politically defined, but may just as easily be social or cultural in inspiration, or more likely a complex combination. Scales emerge from a dialectic of cooperation and competition which always involves social struggle and, as Marston insists, relations of social reproduction are integral with the production of scale (Smith, 1992b). In a broader sense, then, geographical scale is the spatial repository of structured social assumptions about what constitutes normal and abnormal forms of social difference. Scale distils and expresses the oppressive as much as the emancipatory possibilities of space, its deadness as much as its life. The production of scales is contradictory. The generation of scales provides a means of containment, insofar as it rationalizes the identities of specific homes, regions, or nations defined vis-á-vis other homes, regions, and nations, and provides natural territorial bounds for containing specific activities in distinct places. Yet the production of scale is also empowering insofar as it also provides the boundaries of specific places that can be defended in the name of specific identities, social relations or activities.

Some have sought to find an explicit theory of the production of scale in Lefebvre (Brenner, 1997; Brenner, 2000). In various places, Lefebvre gives us tantalizing views of how the production of global space brings about a constant reshuffling and reworking of social spaces at different scales. "Social spaces interpenetrate one another," he emphasizes, and "superimpose themselves upon one another"; spaces may be fractured and differentiated amidst the maelstrom of economic, political and social change, but established spaces may also live on in relatively fixed form. He talks constantly of global and regional spaces, nation-states and cities, all of them open to change, and for Lefebvre, the fate of space is closely bound up with the fate of the state (Lefebvre, 1991). Indeed, long before globalization became such a fashionable question, Lefebvre interrogated the nexus of global space and the state.

While his language of spatial difference is peppered with suggestive comments about scalar change and difference, Lefebvre does not intimate much about what a theory of the production of scale might look like. The discussion of scaled space is not the same as analyzing systematically how space becomes scaled, and there is no systematic discussion of the production of spatial scale. In his sparse theoretical references to scale (as opposed to more numerous empirical discussions of specific scales), Lefebvre often slips between discussions of space in general and scale in particular, and this can be accentuated in secondary discussions of scale that use Lefebvre as a starting point. Further, more traditional treatments of scale still intrude in his fragmented discussions of scale.

As Brenner has pointed out, some of Lefebvre's most explicit statements about scale appear in his four-volume *De l'État* (Lefebvre, 1976). There he extols the multiplicity of scales that comprises a "hierarchical stratified morphology," and is most original vis-à-vis scale in arguing that one needs to know the conditions of "genesis," "stabilization," and "rupture" of different scales if one is to "study them completely" (p. 69). But these highly suggestive comments are made in a specific context, namely Lefebvre's attempt to understand "the globalization ["mondialisation"] of the state," and he is explicit that they comprise "methodological" rather than theoretical arguments (p. 67) (the entire section dealing with scale is entitled "method"). This is clear whenever he talks more generally about the scale question: "The question of scale and of level," he says, "obliges one to choose at the outset the scale one wishes to study" (pp. 67–8). Voicing the crucial contradiction of liberal treatments of scale: "The question of scale, today, appears at the outset . . . of the analysis of texts and the interpretation of events. The results depend on the scale chosen as initial or essential" (p. 68).

With his sense of the "interpenetration" of spaces and the life-cycles (genesis, stabilization, rupture) of scales, Lefebvre begins to glimpse the profundity of the scale question, but his methodological framing and the subordination of scale to questions of globalization and method suggest that he never fully grasped scale as a central theoretical problem *per se* in the analysis of differential space. He passes quickly from method to historically specific discussions of globalization, the state, regions, and cities without the kind of extended philosophical reflection or theoretical ambition that characterizes his treatment of space in *The Production of Space* (1991). Indeed this latter work has surprisingly little to say about scale. The differentiation of scale is not naturalized in Lefebvre – he has moved beyond that point – but the dynamics of scalar restructuring remain largely obscure, and strong threads of the voluntarist treatment of scale linger in his prescriptive comments. Not surprisingly, perhaps, a full realization of the centrality of scale as a metric of spatial differentiation, and a more complete break with the methodological ideologies of scale, had to await the 1980s and 1990s when scale bending events and processes, associated with the restructuring of capitalist regimes and relations of production, were increasingly evident.[2]

After (and Before) Globalization: The Nexus of Capital and the State

"Trade is a bigger prize than ever before in world history," and "world trade has carried us into an era in which *scale* plays an appropriate and highly important part." So announced a well-known American geographer, not in the 1990s when the World Trade Organization rose to prominence, but in the 1940s when such organizations of global governance were just being hatched. The venue was a wartime meeting of the State Department's Political Committee, and Isaiah Bowman was huddled with Secretary of State Cordell Hull and a number of other officials, devising top secret plans for postwar global reconstruction. The purpose of that meeting in June 1943 was to begin the systematic design of what would become the United Nations. Bowman had been entrusted with the job of drafting a first attempt at a UN constitution, and as his five-page effort and subsequent discussions made clear, Roosevelt's government envisaged the United Nations as a political instrument for managing a specifically US-centered globalism.[3]

Not only did Bowman model his UN proposals on the US constitution, replete with universals about human nature and self-evident truths, but he replicated, albeit at a higher scale, the eighteenth-century Federalist Papers debate over the appropriate political geography of the United States. In order to govern such a large territory effectively, Alexander Hamilton, John Jay and James Madison mused, what is the desired balance of centralized (Federal) vis-á-vis decentralized (state) control? In the State Department in 1943, however, it was no longer an issue of constituting a new space at the national scale, but a new space at the global scale. It was in this context that Bowman raised the importance of the "scale question." While some regional prerogative would be necessary and regional differences would have to be managed, he argued, the reality and the prize of world trade inclined them toward a strong global organization. Globalism was preferable to regionalism from the standpoint of US interests, especially if the US could maintain the broad managing hand in global governance that it inherited between 1919 and 1941. US interests flanged neatly with those of other nations, he assumed, since global trade "equalizes the natural resources and advantages of the different regions of the world" (see Smith, 2002: chapter 15).

Bowman's proposal went through many iterations in the State Department and then among the allies, and while the UN charter of 1945 evolved well beyond these origins, the resulting organization and its history bear distinct traces of the politics of scale that framed its early design. Bowman's proposal remains remarkable for a number of reasons, however. First, the application of principles of national construction to the global scale suggests that he and the State Department were involved in some rather ambitious scale bending of their own. Not only were they modeling a global institution after a national blueprint, but they were fudging a specifically national economic interest in world trade as a global interest shared by all. To modify Orwell's concluding dictum of the time, all may be equal in some formal sense but in the world market some are more equal than others.

The directness of the geographer's locution provides a rare glimpse into one strategic episode of scale construction. Although Bowman would hardly have put it in these terms, the wartime US government was embarked on nothing less than a remake of the global scale *as global market*, managed by transnational (largely US identified) capital. The end of World War II was a crucial moment in this process; the period in which an entire infrastructure of global governance was constructed – the UN, The International Bank for Reconstruction and Development (later the World Bank), the International Monetary Fund, the Global Agreement on Trade and Tariffs (GATT, precursor to the WTO), the Universal Declaration of Human Rights, and so forth.

Although it was penned in 1943, the justificatory language of the UN proposal might easily have been written a half century later when the majority leader in the US Senate, Dick Armey, raised free trade to the status of a human right, indeed "perhaps the most fundamental human right." But it might as easily have been written a century earlier as well. The contention that global trade leads to an equalization of conditions across the world – or a flattening of spatial difference – is a staple of globalization ideologies. At least since the 1840s, when the heady expansion of British capital whipped manufacturers into a free trade frenzy, the promise has been that global free trade will deliver commodities more cheaply everywhere, diminishing economic inequalities across the world. This eternal promise of capital-inspired equality at the hands of global free trade has persistently been challenged as a cruel hoax, and never more acerbically than by Marx. Reacting to the repeal of the British Corn Laws, Marx concluded at the beginning of 1848 that free trade was nothing but the "freedom of capital." Far from equality, he argued, global free trade would only increase the antagonism between the owning and working classes. As evidence he cites the early nineteenth-century destruction of the Dacca weaving industry as a result of competition from cheaper machine-made garments in Britain. The confusion of "cosmopolitan exploitation" for a "universal brotherhood" around the world is a grotesque fantasy, he argued. "All the destructive phenomena which unlimited competition gives rise to within one country are reproduced in more gigantic proportions on the world market.... If the free traders cannot understand how one nation can grow rich at the expense of another," he adds, "we need not wonder, since these same gentlemen also refuse to understand how within one country one class can enrich itself at the expense of another" (Marx, 1973: 223, 221; see also Wainwright, Prudham, and Glassman, 2000). Within weeks of Marx's critique of an apparently invincible free trade movement, and as if to prove his point, the economies of Europe convulsed in crisis, unemployment soared, and a series of revolutions broke out across the continent. Implicit in all of this is the same connection between trade and the making of geographical scale that Bowman made a century later. Implicit also is an interpellation of economics and politics in the broadest sense. Then as now, the contradictory geographies of globalization and state formation provide the fulcrum on which many episodes of scale bending balance. There are other sources for scale bending; we could fruitfully have started at a different point in the scale hierarchy, with the body and the home for instance (Marston, 2000). I focus here on the relations between capital accumulation and the state because the shifts taking place at this nexus have been dramatic in recent decades and have reverberated powerfully through other scales.

During periods of enduring economic expansion matched by a stable territorial division between states, assumptions about which kinds of activities fit properly at which scales are also relatively stable. The eruption of scale-bending incidents and events, such as the ones with which I began, suggests on the contrary, a period of scale reorganization in which an inherited territorial structure no longer fulfils the functions for which it was built, develops new functions, or is unable to adapt to new requirements and opportunities. New social activities erode the coherence of old scales and/or crystallize new ones; old activities no longer fit in or support the scaled spaces that hitherto contained them. It is not just that the spatial arrangements of social activity are being reorganized but that the basic territorial building blocks of the social geometry are themselves being restructured. Episodes of scale bending emanate from these deeper shifts. The scale of the nation-state was from the start entangled with questions of capital accumulation. From West Africa to East Asia to Europe, states took many different forms prior to the eighteenth century, each expressing some version of precapitalist social relations: kingdoms and fiefdoms mixed with duchies and city states and so on. New elites constructed nation-states, most commonly by the agglomeration of smaller, previously distinct territories, but also by hiving one territory off from another, by imperial conquest, or by the more gradual morphogenesis of often diffuse kingdoms or precapitalist territories into national units. In many cases it involved all of these processes, as for example with China or Britain. The development of nation-states fundamentally involved two interconnected processes.

First, social definitions of nationality – who comprises "the people" – had to be given territorial definition. Second, the state had to develop a monopoly of violence within the national sphere and defend its borders. These were inextricably interconnected processes and they etched a far more profound historical geography than simply galvanizing an "imagined community."

The means of nation-state formation were occasionally peaceful, more often violent, certainly protracted, but for our purposes the important point is that the main impetus came from the increased scale of economic accumulation in the transition to capitalism and the need to reconcile competition and cooperation, geographical expansion and centralization. This is clearest in the case of Mediterranean Europe where mercantile city states in the fifteenth to seventeenth centuries were economically dependent on a much larger territory than that controlled by the cities, and were simultaneously dependent on non-local armies to defend their investments, sources, and transportation routes. As units of territorial control, city states thus were no longer capable of managing the wider scale conditions of their own economic reproduction. Colonization and economic internationalism were not contradictory but constitutive of national states; imperial conquest was a predicate as much as a result of nation building. Larger kingdoms often fared better in this regard, but equally their capitalist development was retarded by the perpetuation of feudal power rooted in land ownership and absolutist control of the state (Tilly, 1990; Smith, A., 1998; Arrighi, 1994).

The earliest nation-states were the ultimate exercise in scale bending. Spurred on by class revolutions such as those in France and the United States, they combined under one national hat functions that had long been exercised by cities, royal courts, clan chiefs. The problem that nation-states solved was twofold: on the one hand they had to reconcile competition between emerging capitals with the necessity for cooperation in the provision of certain common conditions of social reproduction and production (taxes, roads, labor laws, trade and currency controls, etc.); on the other they had to satisfy revolutionary demands for democratic representation that erupted among peasant and city populations. The protracted worldwide establishment of the national state scale was therefore simultaneously a defensive move by emerging bourgeoisies – vis-à-vis competing capitals and classes – and a progressive one that expanded bourgeois rights into new territories and societies. The nation-state provided a spatial solution – a broad strategy of 'scale jumping' combined with the invention of new institutions – to the combined problems of intraclass competition among capitals and interclass competition for social power. There were precursors of course, such as earlier national states in China or Korea, but as a systematic means of organizing the world's political economy, capitalism and nation-states were born as twins of a dramatically changing seventeenth- and eighteenth-century historical geography in Europe. The scale of the nation-state was defined from the start as container of national capitals, purveyor of national identities, and (in the visage of "national capitals") the elemental building blocks of global political economic competition.

The scale of the nation-state represented a fundamental geographical response to the contradictions engendered by capital accumulation at expanded scale. Capitalism must expand to survive, if Marx's critique is correct, and the emerging system of nation-states provided the political geographical infrastructure to contain and empower that expansion. The jigsaw puzzle of nation-states was always a territorial compromise between global ambition and local control. If the national state system in Europe was largely completed by the Versailles conference in 1919, the global consummation of the national scale had to await decolonisation, which followed much the same dialectic of competition and cooperation, forced by revolutionary demands, the reconciliation of global ambition with local control, and national defense.

The national scale, therefore, represented a platform for a globalization that already preceded and produced it. But this was not a one-dimensional scale expansion so much as a multilateral restructuring of scale. The global reach of capital was possible only because of a parallel centralization of capital at other scales. The centralization of capital in transport or media corporations, in urban and regional economies, in financial institutions, was both a result and a premise of the globalization that accompanied and accomplished nation-state formation.

The fate of the national scale today has to be seen in this light. Globalization per se is not necessarily inimical to nation-state formation; on the contrary the expansion of the world market was historically implicated in the emergence of nation-states. Yet the nexus of global economic expansion and national states looks very different today, and the frenzy of recent bulletins announcing the death of the nation-state – even the end of geography – do catch something real about the current predicament (O'Brien, 1992; Ohmae, 1990; Ohmae,1995; Virilio, 1997; Castells, 1996–8). Such obituaries for the nation-state provide much too simple a prognosis, of course. They express a globalized utopia; literally a globalization that takes us beyond space and spatial difference, in which capital is all powerful, state interference is subdued if not eliminated, and social reproduction is unproblematically guaranteed by the market. The resonance with ideologies of global neoliberalism is unmistakable even in more progressive paeans to the end of the nation-state.

It is widely objected that any erasure of spatial difference is countered by a powerful reassertion of place in the new global geography, And yet this notion of a reassertion of place also does not entirely succeed in grasping the significance of the transformation wrought by the new globalism. The vision from mainstream economic geography is broadly that if the global social economy comprises a plethora of containers – regions and/or nation-states – globalization brings about a dramatic change and resorting of social and economic relations and activities carried on within these containers, and perhaps also an increased porosity of the containers themselves. With the exception of some national containers which could dissolve entirely, the containers themselves remain largely intact even as social relations between and within them are transformed. But the scale arguments of recent years alert us to the fact that with the new globalism, the containers themselves are being fundamentally recast. As Peter Taylor (1994: 159) has put it, "the old wealth containers are no longer operative." Scales are recast and social activities are rescaled (Swyngedouw, 1996). In short, we are witnessing not just the global production and restructuring of space, or of the content of given spaces, but of geographical scale per se.

As regards the national scale specifically, many of the same processes that led to the pupation of national states in the first place now endanger or at least potentially circumscribe the political power that can be wielded at that scale. This *historical* gestalt – similar processes in different periods have diametrically opposite results given different contexts – pivots on the question of scale. While early nation-states provided a means of corralling, managing and rationalizing the expansion of capital, today the scale of capital accumulation has long outgrown the system of national differences that fulfilled these functions. A specific example may help to make this point more concrete.

By the 1970s, the largest automobile companies had ceased producing cars for separate national markets in Europe. New car plants were built with a Europe-wide market in sight. Except for a few cases of specialty cars, it was no longer possible to compete in the automobile market at the national scale. The process was rather different with Japan, East Asia, and Oceania, but the result in scale terms was broadly similar. If car manufacturing presents a highly visible example, the supercession of the national scale was evident in many other markets: shipbuilding, coal, steel, computers, many domestic electronics, to name only a few. The stretching of global space beyond the scale of the national market was matched by an increased global integration and deregulation of financial markets, but these were premised more on the expanded scale of production, and especially the industrial revolution in South, South-East, and East Asia after the 1960s, rather than the other way round. The fiction of a national market could be sustained longest in the US because of the enormous size of the so-called domestic market, but the NAFTA agreement of 1994 solidified what

State Department officials in 1943 well understood, namely that the future of the US economy was necessarily international. They also understood, as globalization afficionados do today, that there is no necessary contradiction between internationalism and nationalism and that the major contest concerns whose national norms – cultural, economic, political – get to become the basis of the new globalism: who gets to be more equal than others in the world market

It is important to note at this point that although my discussion has focused on the global and national scales and on the process of production, the supra-national scale of capital accumulation has dramatic effects at other scales. Scale bending is partly scale stretching but it also implies the fragmentation of pre-existing scales. And it is also intimately connected to the destabilization of identities – national, classed, raced, gendered etc. – in this same period. We will return to this issue below.

At the scale of multinational capitals, there is little doubt that nation-states increasingly represent unfortunate inconveniences on the global map – or, as when they shelter huge reservoirs of cheap labor, conveniences. The assets of Microsoft now exceed the gross domestic product of Spain, and Bill Gates' personal wealth dwarfs that of many poor countries. For them and such as them the notion of national capitals is thoroughly obsolete, and has been for decades. The industrial revolution in Asia since the 1960s together with the heightened mobility of workers and *production* capital in the last three decades has made postnational arguments more real. The post 1980s dismantling of social welfare systems by national states in North America, much of Europe, and Oceania would have been inconceivable without two basic conditions. First, national states have increased their room for political movement as a result of the severe dampening of social protest and of opposition to state policies geared at social reproduction of the national labor force. But second, the definition of a national labor force, much like that of a national capital, is much more porous today with unprecedented levels of global migration and the internationalization of many labor markets. To a greater extent than ever before, many states are freeing themselves from the necessity of local labor force reproduction (Katz, 2002).

Yet, less obviously, many other, especially smaller-scale, capitals still construct and rely on something akin to a national capital and national and local labor markets. Precisely this contradiction concerning scale and power, between a capitalist internationalism and an often nastier bourgeois nationalism, lies at the heart of numerous contemporary debates, and nowhere more so than in the politics of immigration. The representatives of globally mobile capitals are not, by and large, so agitated by immigration into California, France, or Saudi Arabia. Ted Turner and George Soros,

Prince Saud and Bill Gates, whose profits depend on worldwide workforces and on intense yet selective worker mobility, would scorn such pettiness. Smaller capitalists, however, who may benefit from cheaper immigrant labor but who vie for political power at the local and national scales where they are stuck in space, for all intents and purposes, provide the organizing ferment – and much of the financial fodder – for racist anti-immigrant hysteria. The possibilities for a defensive reassertion of national scale prerogatives and privileges in this process are significant and dangerous. Precisely this dilemma is being played out in the United States in the wake of new immigration controls established after September 11.

Nation-states initially crystallized amidst the combined, if contrary, processes of global economic expansion and the centralization of capital, and the new phase of globalization also involves a centralization of capital at different scales. The scale of the nation-state is not automatically weakened but could conceivably be strengthened in certain places as an integral outcome of economic globalization. This could happen from a position of power, to the extent that nation-states are able to insinuate their specific interests as the defining goals of globalization – most obviously the United States in the 1990s – but could also transpire defensively in states that define a national political identity within but against "globalization." This latter perhaps best pertains to several states in Central Africa in the last decade or to Serbia and Croatia and other post-Soviet states (Žižek, 1999).

With appropriate state institutions, therefore, the new globalism can easily generate a new nationalism, but it is equally matched by restructurings of the urban and regional scales, as well as others (MacLeod and Goodwin, 1999; Paasi, 1991; Ohmae, 1995; Scott, 1998; Brenner, 2000; Smith, N., 2000). It is not so much that a "state scale" nestles between global and urban scales (Brenner, 1997: 154–8), but that a thoroughgoing rescaling of the state is occurring as part of this complex set of territorial and political shifts. The notion of a discrete "state scale" invites a confusion between nation and state that obtains in practice during only a short period of history and at the same time encourages the Hegelian slippage that we find in Lefebvre, whereby the state is endowed with a certain teleological impulse to become all. In contrast, it seems to me that the restructuring of scale today, and the rescaling of the state that is integral to but only a part of this process, not only takes place across and throughout the scale hierarchy, from the body to the global, but is fundamentally a process of struggle. The scale question is a lot more complicated than the issue of the globalization of the state. Struggles over the rescaling of the state may be very different in different places, at different times, and at different scales, but a Hegelian teleology of the state does not upset the "logic" of a utopian globalization so much as fit a virtual image – state-based rather than capital-based – into its pre-existing grooves. Either way, the outcomes of globalization and of the rescaling of the state are much more contingent than this conceptualization would seem to suggest.

Conclusion

Vladimir Lenin famously upended Hegel's argument that space eclipses historical time as the state evolves as master of all space. Instead, under socialism, Lenin (1972) argued, the state will wither away. An organ of class oppression, its function fades with the fading of class differences. Lenin's anticipation of the withering away of the state was certainly powered by a sense of agency – a politically mobilized international working class, and carried with it a certain optimism – the world can be made to look very different. Yet fairly or otherwise, Lenin is widely criticized for a certain utopian globalism, and his ambition of a withering state is rarely given voice today except as an object of scorn or nostalgia. The remarkable thing, however, is the virtual reinvention of Lenin's idea at the opposite end of the political spectrum. A left that used to champion the withering away of the state has now evolved, in the context of globalization, into the state's apparent defender, whether buttressed by Hegelian philosophy or liberal sentiment. By contrast, fantasies of the withering away of the state are now the enthusiastic preserve of bankers, financial capitalists, business school professors, and right-wing ideologues preaching free market neo-liberalism and global deregulation. In the 1990s, at least until the Asian economic crisis of 1997–9, Wall Street may have been the real haven of lingering Leninists. Utopian globalism per se is not the issue; the issue is whose utopianism gets to be globalized.

The language of globalization itself represents a very powerful if undeniably partisan attempt to rescale our world vision. Less obviously, perhaps, and at a quite different scale, a similar judgement may well apply to the identity politics which emerged in the 1980s. Identity politics emerged as various 1960s movements – feminism, antiracism, environmentalism, anti-imperialism, and lesbian and gay rights movements among others – developed significant theoretical literatures and installed themselves in the academy. With one foot in the academy and one in activism, these movements mobilized a much wider political instability. Not only were these movements demanding space for the valorization of previously "marginalized" identities, but the economic restructuring of the 1970s was eroding previously stable identities at the same time. Militia movements, antiaffirmative action politics, and the resurgence of right-wing white identity and nationalist groups represent different, partly reactive, reassertions of identity.

Identity politics was inspired by interlocking crises of identity. It represented for some a broadening and deepening of class politics, for others an escape from class. For many it provided a nominally "oppositional" political framework while remaining broadly commensurate with the liberal individualism of North American or European society. Its most radical edge galvanized an ambitious reconquest of the body, a redefinition of the scale of the body, and a dramatic reorganization of the ingredients of identity that go into the making of that scale. This was not happening in a vacuum, of course, but was intimately tied to economic, political, and cultural shifts associated with the emerging new globalism, restructuring of the national scale, and the disruption and remaking of local economies.

Although I have used globalization, capital, and the state as entry points to this discussion of scale in an attempt to sketch some of the processes behind contemporary incidents of scale bending, the emergence of identity politics and the rescaling of the body to which it aspires abets Marston's (2000) observation that theories of the production of scale cannot proceed from a hermetic sealing of relations of production off from relations of social reproduction. The references above to the internationalization of social reproduction, or to the importance of political struggles in the periodic fixing of scales, support this argument. The fate of the national state scale today is unfathomable without a comprehension of the rescaling of social reproduction. A new body politics makes little sense if divorced from the remaking of national and global scales, and vice versa.

It is equally true that the struggles over the reconstruction of scale expressed in scale bending moments are rarely discrete. Not only are different scales interconnected but specific struggles often operate at several scales simultaneously. As one example, consider the "Free Tibet" movement that emerged in the 1990s and has blossomed as part of the antiglobalization politics following the 1999 street battle in Seattle against the World Trade Organization. Devoted to the liberation of Tibet from Chinese tutelage, this movement is variously driven by assertions of national and religious identity, liberal antisocialism, and American protectionism. It combines global, as well as different national and local aspirations, and insofar as it bears on Chinese integration into the global capitalist system (as the appointment of Kissinger as Disney's Chinese ambassador suggests) it impinges on the trajectory of global economic and cultural change. Its central symbols include the berobed bodies of Tibetan monks and likenesses of the Dalai Lama.

This movement has attracted many who champion the local against the global, but more than most such "oppositional" movements in the antiglobalization stable, the Free Tibet movement highlights the scale complexities of globalization. While the demand for human rights may be admirable, this case also highlights, à la Dick Armey, that human rights have become more not less of an economic weapon in the post-Cold War world (Koshy, 1999). On the one hand, an authoritarian religious monarchy is hardly a supportable alternative to global capitalism (or state socialism). On the other hand, support for the Tibetan local against either Chinese authoritarianism or the global free market is widely inspired in the United States by narrow nationalist self-interest, mixed with anti-Chinese racism, aimed at preventing cheap Chinese labor from becoming more directly competitive with US capital. The value of labor power, of course, is a crucial nexus of social production *and* reproduction.

Marx faced the same dilemma, albeit concerning British rather than US-centered globalism: how to oppose free trade while simultaneously eschewing a narrow nationalism. He had no illusions in 1848 that free trade represented anything other than the freedom of capital, nor had he any intention of supporting protectionism. One can declare oneself an enemy of the bourgeois regime, he said, "without declaring oneself a friend of the ancient regime." The "protectionist system is nothing but a means of establishing large-scale industry in any given country" and thereby making it "dependent upon the world market." But the protectionist system "is conservative," he reasoned, "while the free trade system is destructive," he told his Democratic Association audience. "It breaks up old nationalities and pushes the antagonism of the proletariat and the bourgeoisie to the extreme point. In a word, the free trade system hastens social revolution. It is in this revolutionary sense alone, gentlemen, that I vote in favor of free trade" (Marx, 1973: 224).

A century and a half later, that critique of protectionism and insistence on a clear, critical internationalism still provides a fresh alternative to a specifically capitalist globalization. More than anything it suggests that insofar as scales are only ever the temporary spatialization of certain social assumptions, always susceptible to scale bending, the global is every bit as accessible to political struggle as the local. The conquest of scale is a central political goal. Capital may for now make the world in its own image but it does not control the global or any other scale. This is vividly exemplified in the response of the US state after September 11, 2001. Not only did the events of that day have to be anxiously nationalized and the prerogatives of the national state dramatically reaffirmed in order to justify war (Smith, N., 2001), but the "war on terrorism" after October 7 became nothing less than an attempt to secure a specific model of US-inspired and US-led globalism in the one region of the world (Afghanistan and the Middle East) that threatens to opt out of the new globalism – that threatens an "alternative modernity." It is not so much a war against terror as a war *of* state-sponsored terror *for* the global scale. It is a war to secure global political rights to define what does and does not count as terror. As such, that war is not an interruption of 1990s globalization but its continuation by other means, a lesson that the anticapitalist opposition to globalization, mesmerized by the seeming placelessness of global economic and political power, is only slowly learning.

Notes

1 Melissa Gilbert, personal communication.
2 Brenner's dependence on Lefebvre as the source of scale theory thus has the effect of rereading a whole array of theoretical concepts back into Lefebvre's writings after 1968. Lefebvre's movement in this direction was very preliminary, however, and concepts such as "the politics of scale," "rescaling", "the production of scale," "scalar fixes," "scale jumping," and the theoretical arguments about scale that they portend, did not emerge until the 1980s and especially the 1990s. See for example, Taylor (1981); Herod (1991); Swyngedouw (1992b; Swyngedouw, 1996); Smith and Dennis (1987); Smith, (1992a); Leitner and Delaney (1997).
3 "Mr. Bowman's remarks in the political committee, June 12, 1943," Isaiah Bowman Papers, series 52, Special Collections, Milton S. Eisenhower Library, Johns Hopkins University.

References

Arrighi, G. 1994: *The Long Twentieth Century*. London: Verso.
Brenner, N. 1997: Global, fragmented, hierarchical: Henri Lefebvre's geographies of globalization. *Public Culture*, 10: 135–67.
Brenner, N. 2000: The urban question as a scale question: Reflections on Henri Lefebvre, urban theory and the politics of scale. *International Journal of Urban and Regional Research*, 24, 361–78.
Castells, M. 1996–98: *The Information Age: Economy, Society and Culture*. 3 vol. Oxford: Blackwell.
Herod, A. 1991: The production of scale in United States labor relations. *Area*, 23: 82–8.
Katz, C. 1997: Power, space and terror. Unpublished text of lecture, Department of Geography, University of California, Berkeley. October.

Katz, C. 2002: Stuck in place: Children and the globalization of social reproduction. In R. J. Johnston, P. J. Taylor and M. J. Watts (eds), *Geographies of Global Change*, New York: Basil Blackwell, 248–60.

Koshy, S. 1999: From cold war to trade war: Neocolonialism and human rights. *Social Text*, 58:1–32.

Lefebvre, H. 1976: *De L'État. Tome II. De Hegel a Mao par Staline*. Paris: Union Générale d'Éditions.

Lefebvre, H. 1991: *The Production of Space*. Oxford: Blackwell Publishers.

Litner, H. and Delaney, D. (eds) 1997. *Political Geography*, 16(4) Special issue.

Lenin, V. 1972: *The State and Revolution*. Moscow: Progress Publishers.

MacLeod, G. and Goodwin, M. 1999: Space, scale and state strategy: Rethinking urban and regional governance. *Progress in Human Geography*, 23: 513–27.

Marston, S. 2000: The social construction of scale. *Progress in Human Geography*, 24: 219–42.

Marx, K. 1973: On the question of free trade. In K. Marx, *The Poverty of Philosophy*. New York: International Publishers.

Paasi, A. 1991: Deconstructing regions: Notes on the scales of spatial life. *Environment and Planning A*, 23: 239–56.

O'Brien, R. 1992: *Global Financial Integration: The End of Geography*. New York: Royal Institute of International Affairs and the Council on Foreign Relations.

Ohmae, K. 1990: *The Borderless World. Power and Strategy in the Interlinked Economy*. New York: Harper Business.

Ohmae, K. 1995: *The End of the Nation State: The Rise of Regional Economies*. New York: The Free Press.

Quattrochi, D. A. and Goodchild, M. F. (eds) 1997: *Scale in Remote Sensing and GIS*. Boca Raton: Lewis Publishers.

Scott, A. J. 1998: *Regions and the World Economy*. New York; Oxford University Press.

Smith, A. 1998: *Nationalism and Modernity*. London: Routledge.

Smith, N. 1987: Dangers of the empirical turn: Some comments on the CURS initiative. *Antipode*, 19: 59–68.

Smith, N. 1990: *Uneven Development: Nature, Capital and the Production of Space*. Oxford: Basil Blackwell, 2nd edn.

Smith, N. 1992a: Contours of a spatialized politics: Homeless vehicles and the production of geographical space. *Social Text*, 33: 54–81.

Smith, N. 1992b: Geography, difference and the politics of scale. In J. Doherty, E. Graham and M. Malek (eds), *Postmodernism and the Social Sciences*, London: Macmillan, 57–79.

Smith, N. 2000: New globalism, new urbanism: Uneven development in the 21st Century. *Working Papers in Local Government and Democracy*, 99(2): 4–14.

Smith, N. 2001: Scales of terror and the resort to geography: September 11, October 7 *Society and Space*, 19: 631–7.

Smith, N. 2002: *Mapping the American Century: Isaiah Bowman and the Prelude to Globalization*. Berkeley: University of California Press.

Smith, N. and W. Dennis 1987: The restructuring of geographical scale: coalescence and fragmentation of the northern core region. *Economic Geography* 63(2): 160–82.

Swyngedouw, E. 1992a: Territorial organization and the space/technology nexus. *Transactions of the Institute of British Geographers,* 17: 417–33.

Swyngedouw, E. 1992b. "The Mammon Quest": "Glocalisation," interspatial competition and the monetary order: the construction of new scales. In M. Dunford and G. Kafkalas (eds), *Cities and Regions in the New Europe,* London: Bellhaven Press, 39–67.

Swyngedouw, E. 1996: Reconstructing citizenship, the rescaling of the state and the new authoritarianism: Closing the Belgian mines. *Urban Studies,* 33: 1499–1521.

Swyngedouw, E. 1997: Excluding the other: The production of scale and scaled politics. In R. Lee and J. Wills (eds), *Geographies of Economies,* London: Edward Arnold, 167–76.

Taylor, P. 1981: Geographical scale within the world-economy approach. *Review,* 1: 3–11.

Taylor, P. 1994: The state as container: Territoriality in the modern world-system. *Progress in Human Geography,* 18: 151–62.

Tilly, C. (ed.) 1990: *Coercion, Capital and European States.* Oxford: Blackwell.

Virilio, P. 1997: *Open Sky.* London: Verso.

Wainwright, J., Prudham, S. and Glassman, J. 2000: The battles in Seattle: Microgeographies of resistance and the challenge of building alternative futures. *Society and Space,* 18: 5–13.

Žižek, S. 1999: Against the double blackmail. *New Left Review,* 234: 76–82.

10 Is There a Europe of Cities? World Cities and the Limitations of Geographical Scale Analyses

Peter J. Taylor

The advent of globalization as one of the "buzzwords" of our times has provided an unintended fillip to geographical debates about geographical scale. Defined by a geographical scale, the "global," and opposed by other geographical scales, the "regional," the "national" and the "local," it sometimes seems that globalization has finally vindicated two decades of work on scale by geographers. Of course, it is not as simple as this. Whatever globalization is, it is certainly more than simply a bigger organization of society. In this chapter, I am going to emphasize a different aspect of the geography of contemporary globalization, the idea that it is constituted as a global space of flows.

According to Castells (1996), contemporary society is a network society where traditional spaces of places, such as regions and states, are being gradually undermined by new spaces of flows, such as the international financial markets, facilitated by the combined enabling technologies of communications and computers. The world city network is a prime example of such a new space. Typified by a landscape of huge tower blocks of offices, these world cities are connected by a myriad of daily links between these offices across the world. In short, world cities provide an organizational structure to contemporary globalization, a networking of the world. Such a conception is at odds with the usual way of thinking about scale as bordered spaces – how else can you measure the scale of something unless you bound it first? Networks, and the flows upon which they are constructed, can be, and are, constrained by boundaries but the whole point of globalization is the reduction of such restrictions. Thinking in terms of spaces of flows, therefore, frees us from the boundary obsessions of the modern political world map, from narrow "mosaic thinking," and opens up a network approach to geography within which geographical scale is problematic as a core concern(see Leitner, this volume).

As an example, consider the question posed in my title. The empirical sections of this chapter are going to illustrate that indeed there is not a Europe of cities. This may at first seem somewhat disingenuous. Certainly in European Union (EU) circles there is much talk of a Europe of states, a Europe of nations and a Europe of regions, so why not a Europe of cities? Basically my answer is in the nature of the spaces implied by these various terms: like Europe itself, states, nations, and regions define spaces of places, in contrast cities constitute a space of flows. Cities are themselves places, of course, but their relation to space is different. Continents, regions and nation-states constitute mosaic spaces, fully covering the maps that represent them. Cities are different because there are gaps between them. But they are not separate from each other: in those gaps there are flows, of people, commodities, information, and ideas, that connect cities. No city is an island. Cities are the nodes of networks and the flows among world cities today do not greatly respect the boundaries of mosaics at whatever scale, including the EU. Obviously this is all very simplistic when baldly stated – the rest of this paper attempts to explicate this position further – but recognizing this difference is critical for understanding the limitations of geographical scale analyses. Scale is usually associated with boundaries, so that focusing on relations, connections, and networks problematizes geographical scale as traditionally conceived.

The argument proceeds in four stages. First, I will consider types of scale problems as identified in geography and relate them to wider social science concerns. Second, I identify an "embedded statism" as the crucial scale implicated in the very nature of social science knowledge. The result is an analytic "disembedding of cities." I provide nine propositions that link this state of affairs to the neglect of cities as the crossroads of society. This leads, thirdly, to the contemporary literature on world/global cities, where I argue cities are only partially "re-embedded" due to a particular empirical deficit. I provide a new conceptual and empirical framework for world city network studies. Finally, these themes are illustrated by new analyses of European cities in a world of cities.

Aggregations, Levels, Ranges, Spans, Scopes . . . and Geographical Scales

All geographical studies are imbued with issues of scale; choosing scales of analysis, comparing outputs at different scales, and describing constructions of scale are all common practices by geographers. Often these issues were left implicit in a study, but in recent years there has been an explosion of papers treating scale as the object of study in its own right. These studies of geographical scale can be seen as making a fundamental contribution to social sciences as part of its so-called spatial turn. In other words, scale is treated as intrinsic to all social process as, for instance, when global economic restructuring is interpreted as spatial "rescaling." Here I will approach this work from a different direction. I interpret it as representing, to some degree, a "geographicalizing" of problems common to all social science research. In other words issues of "size" and "level" are treated as spatial categories, such as nested regions, so that general problems of analysis or interpretation are presented in specifically areal terms. My starting point therefore, is to set geographical scale in its wider social science context. To do this I delve back into the literature before the latest burst of geographical scale expositions. I identify five basic scale concepts in social science – aggregation, level, range, span, and scope. In brief descriptions of each concept I begin outside geography before proceeding to geographical scale per se.

Statistical aggregations: scale in spatial analysis

Any statistical analysis entails a set of objects upon which measurements have been made to define variables. For instance, examination results provide data for statistical analysis: students are the objects, and their grades for different courses are the variables. Taking any one variable, if the objects are aggregated, the total amount of variance in the variable declines until the aggregation is completed leaving one composite object and thus zero variance. In the school example, there will be a large variation in grades between individual students in, say, geography. The range of differences will lessen when averages for classes are computed, and for the whole year there will be just one overall average geography score (i.e., no variation left after combining all students into a single representative score). Since the pattern of decline in variance will differ between different variables over the same set of objects, however, it follows that the covariance between any pair of variables will alter as the aggregation proceeds. Hence correlations, and all the methods building upon correlations, are potentially unstable as they are dependent on the level of aggregation at which the analysis takes place (Openshaw and Taylor, 1979). For instance, correlations of geography and history grades will differ depending on whether the analysis is carried out at the individual student level or whether classes are used.

This well-known statistical phenomenon is not problematic if the social scientist has a basic theoretical rationale for choosing objects of study, for instance individuals as decision makers in a shopping survey. Geographical research is not so fortunate; what exactly is a "geographical individual" has been long debated (Chapman, 1977). Today we might express doubt as to whether there is such a thing. Often geographers let the data source define the "individual" such as when census tracts are used in urban analyses. Such empiricism is particularly susceptible to the aggregation problem since there are no grounds for choice of scale. This is the "geographical scale problem" as identified by geographers in their quantitative revolution. Changing the scale of analysis produced alternative results meaning that all findings were scale-dependent. There were two reactions to this. Most spatial analysts ignored it, as in the tradition of analysing urban patterns using census tracts. Some analysts attempted to use the scale dependence theoretically, however, to suggest that different covariation at various scales represented diverse processes operating at different scales (Haggett, 1965). Here the lack of a well-defined geographical individual was used positively to provide a flexibility in multiscalar analysis, implying a model of alternative explanations at different levels of analysis. For the development of this model, however, we must look outside quantitative geography to International Relations and its behavioral-quantitative revolution where the "level of analysis problem" was a central issue.

Levels of analysis: systems in political geography

In International Relations (IR) the stimulus to identifying levels of analysis was theoretical. Three levels were identified: the international system, the state, and the individual, although in practice most treatments focused on the first two levels. In recent years the original conception of levels of analysis has been criticized for conflating two different meanings (Buzan, 1995). Are the levels depicted meant to be simply objects of study? This implies they are units of analysis for which measurements are taken, for instance comparing different types of states, or different international orders over time. Alternatively, are the levels actually alternative sources of explanation? This more abstract notion is about identifying at what level we should be looking for the basic mechanisms of international relations, for instance within the domestic politics of states or the balance of power in the overall system. Like IR, a conflation between objects and explanations has been endemic in geography. Without a meaningful notion of geographical individual this conflation was not explored in quantitative geography, but it had catastrophic consequences for political geography.

In the attempts to make political geography a respectable social science the idea of political system was borrowed from political science to provide an ordering frame. When combined, however, with the idea of three distinct scales – international, national, and subnational – the result was to create a political geography consisting of three autonomous systems (e.g., Johnston, 1979; Short, 1982). This went beyond the claim that explanation is scale dependent, to elevate geographical scales to the status of separate ontological entities. This is IR's conflation of object and explanation taken to its geographical extreme. My own intervention was to reconstitute a single system, the capitalist world-economy as an historical system, and to focus upon sources of explanation. The levels of analysis are theoretically integrated as follows: my materialist political geography posits a scale of reality where capital is ultimately realized in the world market, a scale of ideology where the state distorts the market, and a scale of experience where market outcomes are felt in localities (Taylor, 1982). This is to use scale to spatialize ideology, separating experience from reality. Such functionalism, albeit from a very different perspective, has been most developed in spatial economics.

Ranges of goods: central place hierarchies

In the tertiary economic sector, the distribution of service provision has been the subject of normative theory that posits a functional hierarchy of different spatial scales. The act of purchasing goods ranges from routine everyday shopping, for which consumers are willing to travel only short distances, to buying luxury special items for which they will travel long distances. Thus different types of goods elicit different scales of behavior, called the ranges of goods. These different areal ranges of goods combine with the threshold of goods (number of consumers needed to making selling a good viable), to create clusters of like service goods in central places surrounded by their hinterland markets. In this way, ranges and thresholds define spatial scales as overlapping mosaics in the economic landscape.

Central place theory represents a most sophisticated treatment of geographical scales although it has been out of fashion in geography for some time as urban geography has taken its "internalist turn." The service sector continues to grow in relative size in the world economy, however, and services are an integral part of globalization. Hence I will return to cities and hinterlands below.

Spans of control: administrative geographies

Within central place theory there is an "administrative principle" which produces nested hinterlands across scales. Clearly this is a system for bureaucratic neatness and is more generally related to the idea of span of control in classical organizational theory. For an efficient hierarchical organization, a superior can only supervise five, or at most six, subordinates whose work interconnects. Such ideas of administrative efficiency were at the heart of the English reorganization of local government in 1974 (Dearlove, 1979).

Redrawing the local government map as an exercise in organization science meant defining units in terms of optimal population sizes for a given service. In general, there was a policy to equate size with efficiency so that larger units were proposed in order to obtain "economies of scale." This was also presumed to produce higher calibre representatives and officials. And all this theory had to be imposed geographically on very uneven and complicated population patterns. In producing different scales of administrative area for different services, organization theory created an arbitrary hierarchy of spaces of government, often identifiable by their artificial names. Thus organizational "efficiency" far outweighed the idea that administrative geography should reflect the daily lives of the citizenry (Honey, 1981). According to Dearlove (1979), this was an attempt to control public power with an abstract scaled neatness replacing messy local conflict.

Scopes of conflict: production of geographical scale

Bringing power into consideration, local government reform can be viewed as an exercise in defining the scope of conflict. Power accrues to those who define the scope of a conflict according to the seminal political analysis of Schattschneider (1960). By controlling who is in a conflict and who is outside, the relative balance of strength between combatants is defined and the outcome thus determined. American federal politics has a long tradition of limiting the scope of conflicts to contain radical challenges to established power. A basic means to contain conflict in this sense is geographical through defining the scale at which disputes are resolved and decisions made (Taylor, 1984).

This traditional political science approach relates to the main recent theorization of scale in human geography: the production of geographical scale. This social constructionist approach argues that there are no "natural scales," the assignment of activities to specific scales is historically contingent upon power forces that create scales amenable to their ends. Thus contemporary economic restructuring can be viewed as a conscious "rescaling" of markets and regulations both "upwards" and "downwards" away from the nation-state (e.g., Brenner, 1999; Swyngedouw, this volume).

This "glocalization" has brought forth a new dyad of scales and lessened the traditional dominance of the state as the "natural" scale of political activities and much else besides.

Aggregations, levels, ranges, spans, and scopes are all social science conceptions of relative size, whose arguments have been mimicked in discussions of geographical scale. In what follows I am going to put some of these different ways of thinking about scale together in a relational argument that moves us from an areal focus to a network one. In other words, using another social science perspective (Castells, 1996), what happens to relative size and geographical scale when we think in terms of a space of flows instead of a space of places?

Embedded Statism and Disembedding Cities

This section summarizes arguments that I have made elsewhere. Here they are brought together as a sequence of propositions that begins with states and finishes with cities. I provide just a minimal justification for each statement, referring to other more detailed discussions. My purpose is simply to provide the intellectual context for the empirical contribution in the next section, where I return to Europe and its cities.

PROPOSITION 1. *The nation-state defines the pivotal scale in geographical scale analyses.* This is reflected often in the language used to express other scales such as "international" and subnational." Where the language is not explicit, as in global-local nexus or glocalization, the importance of the concept relies on its omission of the state: even in its silence the state remains pivotal. Our thinking about geographical scale has itself been state-dependent (Taylor, 1982).

PROPOSITION 2. *"Society" is normally conceived as a people geographically bounded by the sovereign limits of a nation-state.* This is part of a remarkable spatial congruence assumption where society, economy and polity are deemed to match state territory. These concepts lie at the heart of our orthodox social theories, creating a critical lacuna at the heart of the social sciences. This unexamined geographical discourse has meant that core social science disciplines – economics, political science, and sociology – have been unconsciously state-dependent. Social analyses have been fundamentally state-centric in nature, privileging one geographical scale over all others (Taylor, 1996; Taylor, 1997a).

PROPOSITION 3. *There is a "states-metageography" that dominates the way the world is routinely perceived at the global scale.* The world political map of states, found in every geography school room, is by far the most familiar of all human geography maps. It is sometimes said the lines on the map depicting state boundaries appear to be as natural as rivers and coasts. In this case state-dependence is expressed as a mosaic structure of space: a space of places (Taylor, 1994; Taylor, 1995).

PROPOSITION 4. *States are the great producers of publically accessible evidence on all social activities, which is why such data are called stat-istics without the hyphen.* Our social world is described by states for state purposes. Whether UN statistics at the international scale or local statistics at the subnational (regional or urban) scale, state agendas are intrinsic to the information gathered (Taylor, 1996; Taylor, 1997b).

PROPOSITION 5. *As territorial political entities, states are largely interested in "taking stock," counting their populations and resources in state-defined spaces.* This means that the vast majority of publically accessible data provides attribute measures of areas, to the relative neglect of relations, connections and flows. We are provided with information on a space of places, but not on a space of flows (Taylor, 1999).

PROPOSITION 6. *Although connectivity is the raison d'être of cities, there is very little data on relations between cities.* State census volumes focus on cities as places in which to take stock, not as the crossroads of society. Hence although there are numerous references to relational concepts such as city networks, urban hierarchies and city systems in the literature, the evidential basis documenting these ideas has been limited. Cities are "denetworked" in statistics (Taylor, 1996; Taylor, 1997b).

PROPOSITION 7. *At times cities have been lost in social analyses that focus upon "urbanization," a stock-taking areal concept eminently suited to statistical description.* To view the growth of cities in a simple areal manner is the ultimate degradation of the city as a crossroads in a wider world. Instead it is represented as an essentially local problem, the "invader" of rural idylls. We need to think, in Murray Bookchin's (1995) terms, of "cities against urbanization": statistics have promoted anti-city ideologies (Taylor, 2000a).

PROPOSITION 8. *Where cities are conceptualized in relational terms they have, until recently, been geographically truncated as "national urban systems."* This nationalization of cities has removed connections beyond the state-space from the city-system analysis. Theorizing urban systems on a state by state basis as "primate-city" distributions, that develop into "rank-size city systems," totally downgraded, if not wilfully ignored, relations beyond the nation-state boundary. Cities and their relations should only be truncated in this way under absolutely autarkic state conditions (Taylor, 1996; Taylor, 2000a).

PROPOSITION 9. *Studies of a world city network as an alternative meta-geography confront a veritable evidential desert.* The evidential structure of the world cities literature includes very little on intercity relations. Data on relations between world cities suffers from a double deficiency: a state-centric data bias complemented by an attributional measurement proclivity. Despite much reference to networks, hierarchies and systems of cities at a global scale, in fact we know little about how world cities connect with one another (Taylor, 1999, 2000a).

World Cities as Networks in a Network of World Cities

The title of this section is unashamedly taken from Brian Berry's (1964) seminal paper "Cities as systems within systems of cities." I have "globalized" the original and use it as an idealized starting point for considering contemporary world cities. It might seem odd to hark back to the 1960s in this context, but relations between cities were less neglected in the urban geography of the quantitative revolution compared to most current research on world cities. Certainly they were nationally truncated but they were still treated as parts of networks. The "internalist turn" in urban geography (Taylor, 2001a) has meant that relations within cities have far out-weighed concern for relations between cities over the last few decades – Berry and Horton's (1970) famous urban geography text is the last to seriously attempt to provide a balance between the internal and external relations of cities. Derivation of my title from Berry is thus an explicit signal of my intention to return to his balance.

Bringing the external relations of cities back into focus also allows me to build upon foundations laid a generation ago by two of the University of Minnesota's most famous geographers, John R. Borchert and Fred Lukermann. Borchert (1967) bemoaned the relative neglect of research on the development of systems of cities compared to individual cities and set out to right this omission in his seminal study of US city systems over four epochs from 1820 to the 1960s. Lukermann (1966) made the critical point that nodality and hierarchy cannot be understood simply by enumerating populations and functions; measures of flows, exchanges, connections or relations are also needed. In what follows I will derive lessons from both these Minnesota geographers for the study of contemporary world cities, starting with the latter.

A lesson from Lukermann: beyond the world city hierarchy

It seems to be accepted that it is in the nature of cities and towns that they operate at different geographical scales to form urban hierarchies. That's what central place theory says and studies of national urban systems have confirmed this truism. For instance, the diffusion of innovations was modeled as a hierarchical process from national metropolis through other metropolises to regional and local cities (e.g., Bourne, 1975). When world cities were identified in the 1980s, the structure of their interrelations was therefore treated as a given: there must be a world city hierarchy (Friedmann, 1986).

If we have learned nothing else from geographical scale analysis, however, we should be suspicious of research that simply moves processes from one scale to another. Changing scales means constructing a different set of social relations, so moving up from a national to a world hierarchy model must at least be problematized. This has not been the case. Beginning with John Friedmann, the ordering of world cities into a hierarchy has dominated world city thinking, despite a paucity of evidence. This has been subsequently augmented by Roberto Camagni's (1993) famous diagram of "the hierarchy of city networks," with world cities at the top, although again no strong evidence is provided. The problem with these structural conceptions is illustrated in Peter Dicken's (1999) *Global Shift*, a key textbook whose value lies to a large extent in its bringing together reams of evidence to describe contemporary transformations in the world economy. There is no such evidence when he turns to world cities. In his diagram that seeks "to give an impression of a connected network of cities" (p. 209) we are told that "the links shown are diagrammatic only." Looking at just the European part of the diagram (Figure 10.1), it is not at all clear why the link between Dusseldorf and London goes through first Brussels and then Paris. Why such an indirect, three-step connection in this electronic communication age? This example illustrates both the poverty of data, given the need to resort to "diagrammatic links," and the assumption that there must be some hierarchical structure in a distribution of cities.

According to Lukermann (1966), defining a hierarchy requires more than just counting, or producing attribute measures. For a hierarchy to exist there has to be some notion of control up and down different levels: "each hierarchical level has autonomy over orders below itself, while being dependent on those above" (p. 17). In other words there has to be evidence of "a *line* of command" (p. 18, Lukermann's emphasis). World city studies are full of rankings of cities supposedly showing a hierarchy. Lukermann's lesson is that available information is totally inadequate for this task: to rank is merely to order by a size measure, it need have no relation to hierarchical structure (Taylor, 1997b). Friedmann (1986) did indeed allocate "command and control" functions to his world cities in the form of locations of corporate headquarters in his initial work that produced the world city hierarchy. In this situation, however, it is not the cities that are

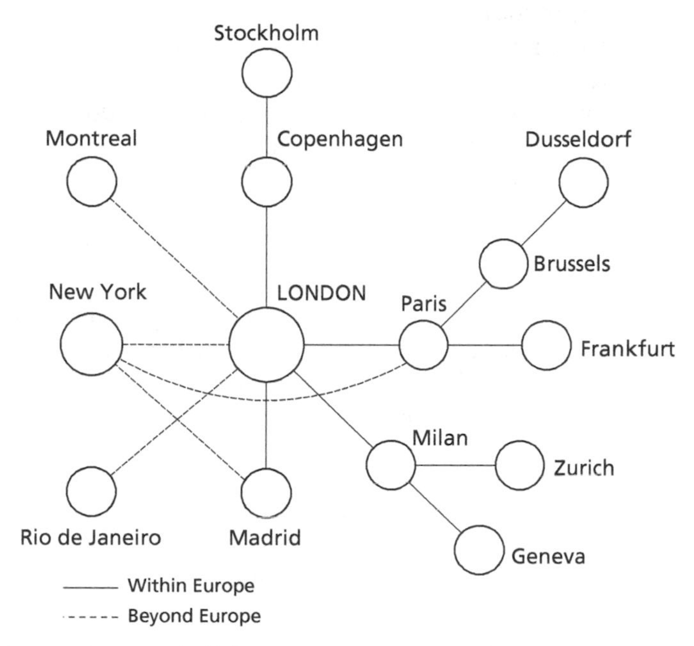

Figure 10.1 *"Diagramatic links" among European cities.*
Source: Part of Figure 7.4 in Dicken (1999).

doing the commanding – London telling Paris what to do – but rather the hierarchies are in the firms. This is much more complicated than a simple world city hierarchy structure (Taylor, 2001b).

Specifying intercity relations at the global scale

There are many ways we can begin to specify relations between world cities. The most common is to map infrastructure patterns, notably airline routes. These may provide good general pictures of world city links but they are best seen as enabling mechanisms rather than the particular social relations that define a world city network. The latter have been identified as deriving from the emergence of advanced producer services as a cutting edge of the new globalizing economy (Sassen, 2000). Cities have always been service centers, of course, but in the new circumstances of economic globalization particular services have had to be produced to satisfy complex and original needs emanating from intense multijurisdictional activities. Whether it is new financial products or interjurisdictional commercial law, service firms have grown to provide necessary inputs to make operating in a global market feasible and profitable. To carry out this task, service corporations need to locate in places where there are large stocks of knowledge continually augmented by flows of new information. These are world cities, where face-to-face communications facilitate economic reflexivities in dense local clusters of service practitioners (Sassen, 2000).

This is the world city not as the home to all kinds of multinational corporations but rather as the locus of one particular type: the producer service providers – accountants, realtors, advertisers, bankers and financiers, lawyers, management consultants, insurers, and so on. Located together as knowledge clusters, this is the world city as itself a network (Storper, 1999). But what of the relations between cities? The firms providing these services cannot do their job from just one city. Tying together deals across several jurisdictions requires being located in many cities, each with its own distinctive special knowledge mix. Hence global service firms have networks of offices across world cities, to provide a seamless service for clients across all, or most, parts of the world economy. World cities are therefore nodes in information flows within global service firms, which link together knowledge clusters to create unique products for global clients (Taylor, 2001b). Figure 10.2 illustrates a miniscule section of this network, for ten cities and three firms[1].

At GaWC[2] we have focused upon the office networks of service firms as the most promising window into intercity relations. As well as having a theoretical rationale, this is also an area where data can be collected reasonably easily to fill the empirical void in intercity studies, and for world cities in particular. We have constructed a unique data set for 46 global service firms, defined as having offices in at least 15 different cities. This covers more than 250 cities across the world although we focus below on those we define as world cities, 55 in all, plus other European cities for which we have found some evidence of world city formation (see Beaverstock, Smith, and Taylor, 1999).

A lesson from Borchert: from hinterlands to hinterworlds

An urban hierarchy assumes that cities with different sized hinterlands provide different levels of service. In his tracing of the development of the US urban system Borchert (1967) noted that rates of city growth were scale dependent, varying by the size and resource base of a city's hinterland. This scale of hinterlands changed over time in response to changing technology, especially in transport: Borchert illustrates how transport technology directly affects the size of hinterlands. Thus it is hardly surprising that his four epochs are each distinguished by a major change in transport technology. He illustrates this as a changing urban hierarchy from "sail and wagon" through to the internal combustion engine. Since Borchert's work, we have

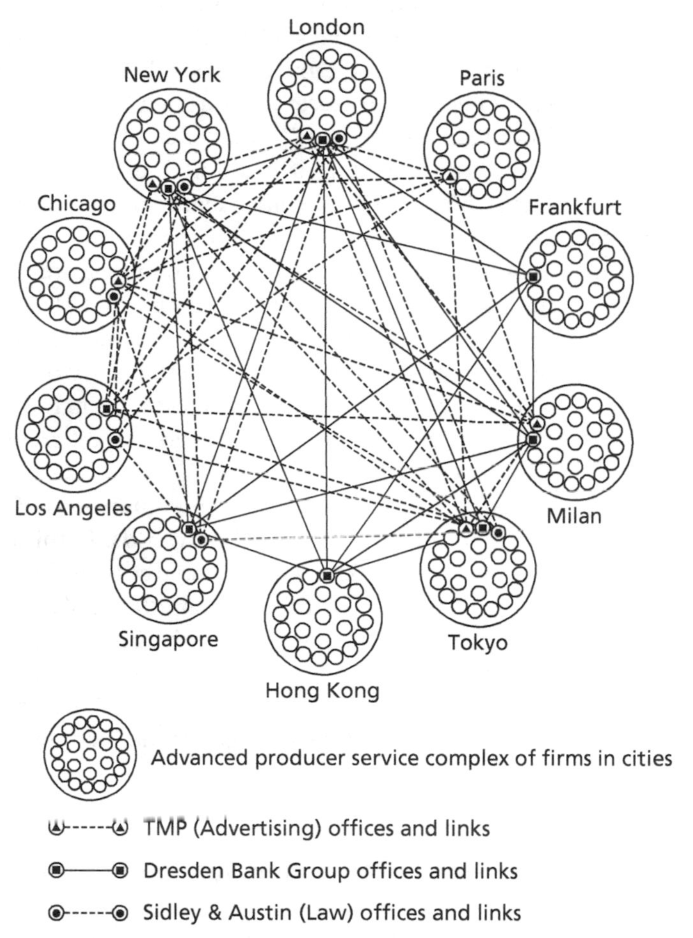

Figure 10.2 *A miniscule section of the world city network as an interlocking network: ten "alpha" cities and three advanced producer service firms.*
Source: From Taylor (2000b).

clearly entered another technological epoch in his terms: electronic communication is revolutionalizing the delivery of services in many different ways. As he states: "unforeseen major changes affect old cities and new alike" (p. 331) so we need to think afresh for every epoch. This is particularly true for the current epoch with its instantaneous global communication capability. Although this facility has existed for some time now, it is not clear that we have got to the stage of revising our concepts to accommodate this remarkable new circumstance. I will suggest one such change.

Ideas derived from central place theory lead us to expect that cities have distinctive hinterlands, and that the cities themselves define an urban hierarchy. But with electronic communication and global suites of offices for large service firms it is not at all clear that such thinking is any longer relevant for many important business services. Doing business between Minneapolis and Melbourne does not need any intercession through Chicago or Sydney, the respective neighboring "higher level" cities. Seamless service can be provided directly, if the firm providing it has offices in both Minneapolis and Melbourne. In fact all 55 of our world cities share some of the 46 firms in common, making it possible to commission direct servicing for all dyads. Obviously it varies between cities – New York and London share more firms with more cities – but this does not take away the point that all world cities have global capabilities. Hence, instead of thinking about hinterlands for world cities we must consider hinterworlds. The latter are worldwide and therefore not bounded. The way they differ between cities is in the pattern of service, the amount of seamless service that is possible in different parts of the world (i.e., through other world cities).

Taylor (2001a) discusses how to measure varying intensities of seamless service capabilities and maps of different city hinterworlds. In this case what is vital to a city is the information/knowledge resources within its hinterworld, and not the physical resources within the hinterland. We can say that behind every successful world city there is an intensive hinterworld. But the key point here is that hinterworlds are all the same in terms of scale whereas hinterlands vary by scale depending on the size of the city. Rather than being areal in nature like hinterlands, hinterworlds are defined by a service delivered through the world city network.

European Cities in an Interlocking World City Network

In contemporary scale analyses, levels above the state are usually identified as regional and global, with the former nesting into the latter. Europe, as represented by the EU, is the regional scale with the most advanced supranational institutions. All the previous discussion implies that whereas political institutions combining states can define a bounded place, they are much less relevant to spaces of flows. Since cities under globalization define nodes in a world city network, it is not at all clear how an institution such as the EU can accommodate the cities located within its realm with their many extra-European connections. These may be "European cities" on account of their location, but it is highly unlikely that they constitute a European network or hierarchy of cities. This can be investigated empirically.

The GaWC data define 53 cities in Europe as world cities or as having evidence of world city formation processes (Beaverstock, Smith, and Taylor, 1999). Each of these cities has its particular mix of global service firm offices, the result of locational decision making by service firms when they decide where to make the large investment of setting up an office in a world city. The resulting office, of course, has to be staffed by professional practitioners, who add to the knowledge concentration in the chosen world city.

I assume that, in their striving for a seamless service, global service firms' office networks represent fundamental distributions of flows of information and knowledge between cities. Thus it follows that two cities with similar service complexes will likely have more commercial relations than two cities with very dissimilar service complexes. This assumption has not yet been empirically tested, but it is a plausible enough starting point in this particular research area, intercity relations, with such deficit of evidence.

Parsimony and components

The assumption about similar and dissimilar service complexes suggests the need for a study of covariances between cities. Every one of the 53 European cities can be correlated with the others in terms of service firm provisions, creating 1,378 measures of relations between pairs of cities. These define the total covariation in the data. The resulting correlation matrix can be simplified using a standard principal components analysis. This is an exercise in parsimony that is very necessary given the complexity of the situation faced here.

The basic output of a principal components analysis is as follows:

1 a reduction in the number of dimensions of variability – cities are replaced by a relatively small number of composite dimensions of variability, or "components";
2 each original city correlates, or "loads," with each of the components – normally a city's "loading" (correlation) on one component is much higher than all its other loadings;
3 from the pattern of loadings, components can be interpreted in terms of which cities load highest;
4 this produces a clustering or classification of cities by their service complexes;
5 because cities load on all components, a strict classification is not produced, since some cities will load relatively high on more than one component, so overlapping groups of cities are created;

6 each city's component loadings produce a measure of "communality" which shows how much of that city's variability is described by the principal components analysis;
7 from city communalities the total variation in the data accounted for by the components analysis can be computed.

For details of this methodology and its application to the GaWC data, see Taylor and Walker (2001).

A European-scale analysis

The 53 European city service complexes are reduced to just 5 components that account for 69 percent of the original variance. Although there is no logical reason for these results to define a spatial order, this is the case here (Taylor and Hoyler, 2000). Globalization does not simply brush aside history, but builds upon past processes in creating new geographies. In this case the oft-identified "city-studded" north-south "spine of Europe," whose origins going back to the earliest modern developments (Rokkan, 1970) is clearly reproduced as the spatial order in this analysis of contemporary decision-making by global firms. Figure 10.3 shows this order as five types of service complex, which fall into two spatial groups. First, there are the cities of the central spine of Europe: (i) minor spinal cities such as Arhus (ARH), Hamburg (HAM), and Genoa (GOA, and (ii) major spinal cities such as London (LON), Paris (PAR), and Milan (MIL). Second, there are the more peripherally located cities, which fall into three types: (iii) cities of the outer triangle such as Oslo (OSL), Lisbon (LIS) and Budapest (BUD), (iv) far west (British Isles) cities such as Dublin (DUB), Edinburgh (EDI), and Leeds (LEE), and (v) far east (ex-Soviet bloc) cities such as Moscow (MOS), Warsaw (WAR), and Kiev (KIE). Each of these types has a common mix of global service offices. Note the hierarchical tendency, where the core spine of European cities is divided between two levels of city. One notable feature of these results is that London is the only city in the British Isles not to load on the British factor. It seems that London is not very British in its service complex mix.

This is not the whole story, however. Table 10.1 shows the cities with the lowest communalities. I will call these "un-European" because they have complex mixes with less similarities to other European cities. The eight cities listed here fall into two groups, eastern European, and more interestingly, cities with important transnational functions. London is not only un-British, it is also un-European. The implication is clear and hardly surprising: we cannot capture the essence of London's corporate mix in an analysis at the

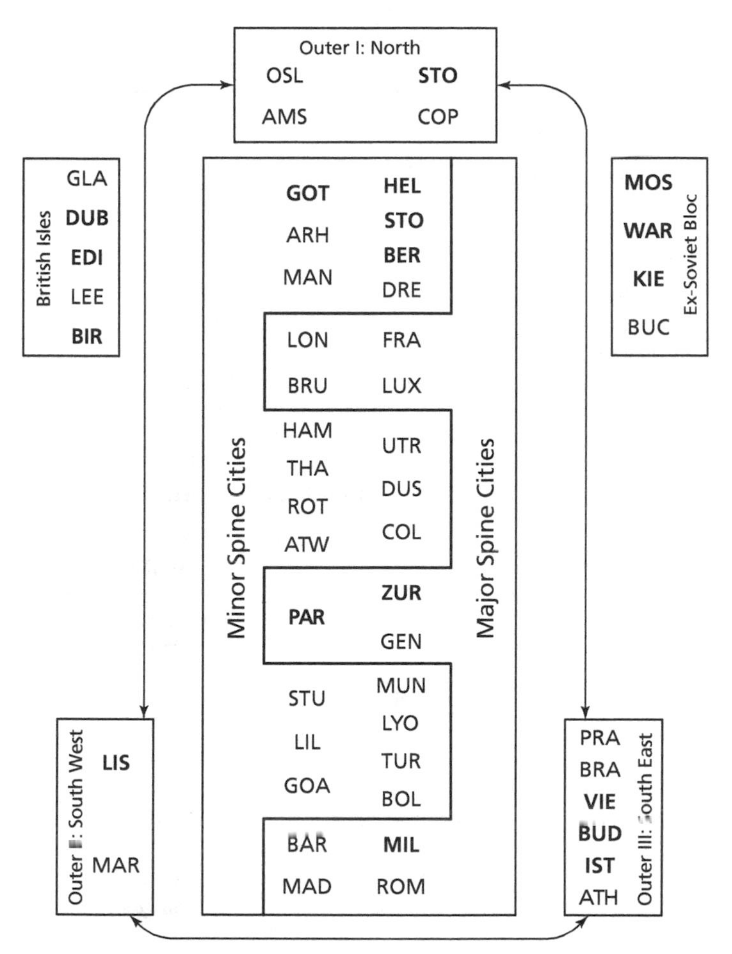

Figure 10.3 *The spatial order of European cities. Cities are allocated by their highest loading. Emboldened cities do not have loadings above 0.4 on any other component. ATH = Athens, AMS = Amsterdam, ARH = Arnhem, ATW = Antwerp, BAR = Barcelona, BER = Berlin, BIR = Birmingham, BOL = Bologna, BRA = Bratislava, BRU = Brussels, BUC = Bucharest, BUD = Budapest, COL = Cologne, COP = Copenhagen, DRE = Dresden, DUB = Dublin, DUS = Dusseldorf, EDI = Edinburgh, FRA = Frankfurt, GEN = Geneva, GLA = Glasgow, GOA = Genoa, GOT = Gotenburg, HAM = Hamburg, HEL = Helsinki, IST = Istanbul, KIE = Kiev, LEE = Leeds, LIL = Lille, LIS = Lisbon, LON = London, LUX = Luxembourg, LYO = Lyon, MAD = Madrid, MAN = Manheim, MAR = Marseille, MIL = Milan, MOS = Moscow, MUN = Munich, OSL = Oslo, PAR = Paris, PRA = Prague, ROM = Rome, ROT = Rotterdam, STO = Stockholm, STU = Stuttgart, THA = The Hague, TUR = Turin, UTR = Utrecht, VIE = Vienna, WAR = Warsaw, ZUR = Zurich,*

Source: From Taylor and Hoyler (2000).

Table 10.1 *The most "un-European" cities*

City	*Communality*
Budapest	0.432
St Petersburg	0.433
London	0.456
Helsinki	0.482
Istanbul	0.533
Zurich	0.543
Paris	0.568
Luxembourg	0.571

Source: From Taylor and Hoyler (2000)
Note: Communalities show how much of a city's variance is accounted for by the five components

Table 10.2 *Major European "spine cities" in the global-level analysis*

Spine city	*Global-level city cluster*	*Non-European example*
London	Global City	New York
Frankfurt	Transnational	Tokyo
Brussels	Western Europe	
Paris	Western Europe	
Zurich	Transnational	São Paulo
Geneva	Transnational	Sydney
Barcelona	Western Europe	
Milan	Transnational	Toronto
Madrid	Transnational	Singapore
Rome	Eastern Europe	

European scale. London is a global city and can only be understood fully at that scale. In other words, even though the analysis is at a world regional scale, London is being truncated just like the national analyses reported previously. The same is true to varying degrees for all the major transnational European cities with their important extra-European linkages.

A global scale analysis

What happens to these European cities when cities from the rest of the world are brought into the analysis? The 55 world cities includes 22 European world cities from the previous analysis (Taylor and Walker, 2001). Using the same methodology, nine components were identified as service complex types that account for 80 percent of the initial variation in the data.

The pattern of cities in this global-level typology is much more complex than at the European scale, as we might expect (Figure 10.4). Although the results are quite regional in nature indicating a spatial patterning to service complex types, European cities feature in a very fragmented way. There are three minor components which are clearly European, one for Western Europe and two for Eastern Europe, but they involve only a few European cities. Northern European cities, particularly German ones, combine with smaller North American cities to create a "North Atlantic" cluster. Many major transnational centers such as Frankfurt (FRA), Milan (MIL), and Madrid (MAD) are part of another major cluster with Tokyo (TOK), Sydney (SYD), and Latin American cities. London (LON) is separate again: in this analysis London is able to quite appropriately join with New York (NY) as a global city cluster whose links are more with other US cities than with Europe.

In other words, European cities soon become dispersed once the rest of the world is brought into the analysis. Focusing on the "major spine cities" from the earlier analysis, Table 10.2 shows how they are distributed among service complex types in the global analysis. Of the ten cities, half are to be found in the transnational cluster with key links far beyond Europe. This is the geographical reality behind the complexity of globalization: a world city network with both regional and hierarchical tendencies.[3] There seems to be the clear answer to my title question after these analyses. Quite simply, you cannot contain world cities within bounded spaces of any scale. Cities by their nature are networked and their analysis should not be truncated. Even at as large a geographical scale as Europe, indigenous cities cannot be contained under conditions of contemporary globalization.

Conclusion: Scale and Network

Returning to the question posed in my title, there is no Europe of cities. Quite simply, world cities cannot be contained within bounded spaces of any scale. Cities by their nature are networked, and their analysis should not be truncated. Even at as large a scale as Europe, indigenous cities cannot be contained under conditions of contemporary globalization. I identify from this four implications for studying geographical scale.

First, in terms of globalization, two general approaches can be found in the literature: one that emphasizes globalization as a special scale, and another that emphasizes the new intensities of transnational flows. Clearly globalization is both of these but there remains the question of balance

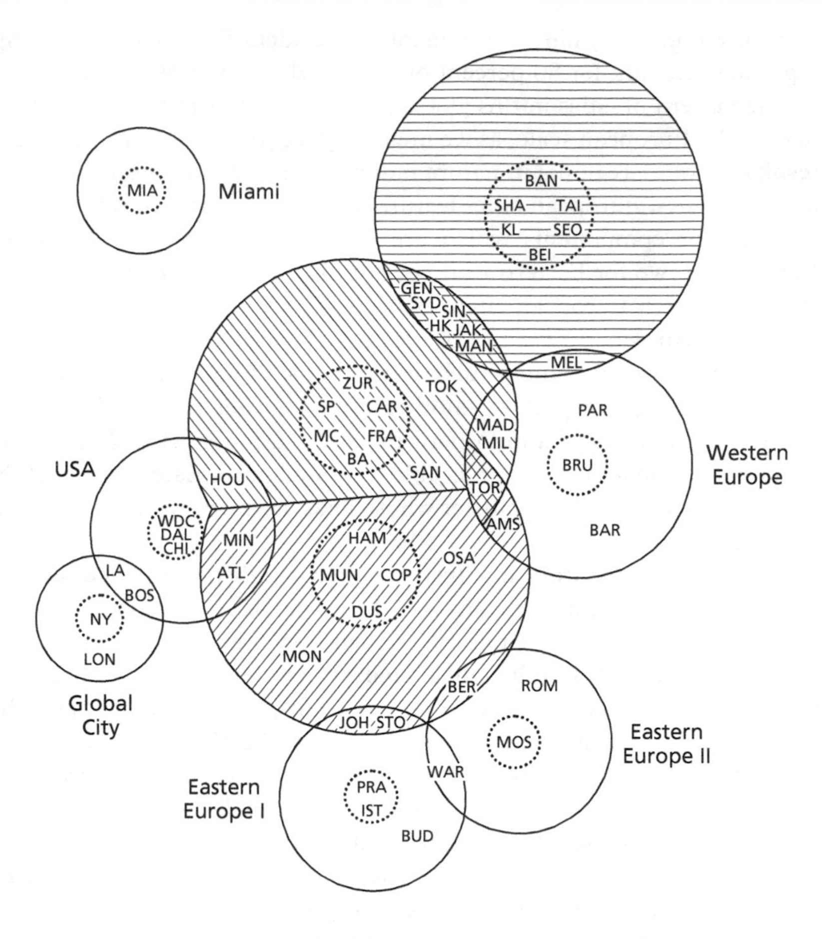

Transnational and Latin American world cities

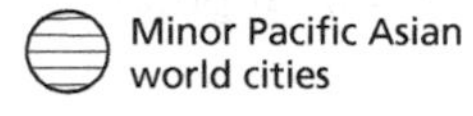

Minor North Atlantic world cities

Figure 10.4 *Hierarchical and regional tendencies in world city service complexes. AMS = Amsterdam, ATL = Atlanta, BAN = Bangkok, BAR = Barcelona, BEI = Beijing, BER = Berlin, BOS = Boston, BRU = Brussels, BUD = Budapest, BA = Buenos Aires, CAR = Caracas, CHI = Chicago, COP = Copenhagen, DAL = Dallas, DUS = Dusseldorf, FRA = Frankfurt, GEN = Geneva, HAM = Hamburg, HK = Hong Kong, HOU = Houston, IST = Istanbul, JAK = Jakarta, JOH = Johannesburg, KL = Kuala Lumpur, LA = Los Angeles, LON = London, MAD = Madrid, MAN = Manila, MEL = Melbourne, MC = Mexico City, MIA = Miami, MIL = Milan, MIN = Minneapolis, MON = Montreal, MOS = Moscow, MUN = Munich, NY = New York, OSA = Osaka, PAR = Paris, PRA = Prague, ROM = Rome, SAN = San Francisco, SEO = Seoul, SHA = Shanghai, SIN = Singapore, SP = São Paulo STO = Stockholm, SYD = Sydney, TAI = Taipei, TOK = Tokyo, TOR = Toronto, WAR = Warsaw, WDC = Washington, DC, ZUR = Zurich*

Source: From Taylor and Walker (2001).

between them. The territorial tradition within embedded statism as practiced in the core social sciences and human geography leads to a privileging of scale, the world as the largest social space of places. I argue that this situation, of an "erosion of the state," represents an opportunity to reassess the importance of the space of flows (Taylor, 2000b). The social sciences are ripe for seeking a new balance between attribute and relation, between places and flows, that will problematize scale as simply territorial size.

Second, analyses of world city complexes show a very complicated geography of globalization that is highly regional in nature. This does not mean a return to mosaic thinking and scales, however. The ordering within the network is no simple territorial pattern: there is much interpenetration of spaces so that no contiguous configuration of regions occurs. The resulting spatial orders represent subnetworks of cities, groups of cities, often part of the same world region, which share similar mixes of service firms. As part of a larger whole, a subnetwork is, of course, not a bounded separate entity.

Third, in terms of the relations of geographical scale to different social science conceptions of size, it is clearly the levels of analysis approach that is implicated in the study of world city networks. The specification of the world city network defines three levels of analysis in a systemic framework: inter city relations as network, world cities as nodes, and global service firms as subnodal actors. Although the quantitative analysis treats these levels essentially as "objects," the ultimate interest will be in defining different sources of explanation through these levels. It is important to note that the specification privileges the subnodal firms as the prime level at which to start searching for explanations. The analyses do not define a top down scale process.

Fourth, the production of a global scale has been the work of many agents and here we have focused upon just one set, global service firms. A key point, of course, is that their production has created a single global network not a single "global region." Furthermore, this global scale is produced in cities and through cities by firms. But they are not the only agents in this particular scale production. Production of any scale is ultimately political and this is true of the global. Clearly it is necessary to bring political geography back into consideration. What is the political geography of a network? What is the political geography linking spaces of places to spaces of flows (Taylor, 2000a)? These questions impinge directly upon the nature of a modern politics that has been very territorialist in nature and organized largely by geographical scale (from local township/commune/parish to nation-state and above). Moving from territorialist alliances at different scales to network-based alliances, such as leagues of cities, produces a different (scale-free?) politics and a new political geography we have hardly begun to acknowledge, let alone study (Leitner, this volume).

Notes

1 This specification is spelt out in detail in Taylor (2001b). World cities constitute an interlocking type of network, which is relatively unusual in networks of social relations. The implication is that the key "actors" are not the city nodes but rather the "subnodal" firms.
2 GaWC is the Globalization and World Cities Study Group and Network organized from Loughborough University and available at http://www.lboro.ac.uk/departments/gy/research/gawc.html.
3 A Nystuen and Dacey (1961) graph theory analysis reveals little of interest. Using 55 cities just two levels emerge: London/New York and the rest. In fact this technique automatically "finds" a hierarchy whatever the structure of the data.

References

Beaverstock, J. V., Smith, R. G. and Taylor, P. J. 1999: A roster of world cities. *Cities*, 16, 445–58.

Beaverstock, J. V., Smith, R. G. and Taylor, P. J. 2000: World city network: A new meta-geography? *Annals of the Association of American Geographers*, 90: 123–34.

Berry, B. J. L. 1964: Cities as systems within systems of cities. *Papers of the Regional Science Association*, 13: 147–65.

Berry, B. J. L. and Horton, F. E. 1970: *Geographic Perspectives on Urban Systems*. Englewood Cliffs, NJ: Prentice-Hall.

Bookchin, M. 1995: *From Urbanization to Cities*. New York: Cassell.

Borchert, J. R. 1967: American metropolitan evolution. *Geographical Review*, 58: 301–23.

Bourne, L. S. (ed.) 1975: *Urban Systems*. Oxford: Clarendon Press.

Brenner, N. 1999: Globalization as reterritorialization: The rescaling of urban governance in the European Union. *Urban Studies*, 36: 431–52.

Buzan, B. 1995: The level of analysis problem in international relations reconsidered. In K. Booth and S. Smith (eds), *International Relations Theory Today*, Cambridge: Polity, 198–216

Camagni, R. P. 1993: From city hierarchy to city network: Reflections about an emerging paradigm. In T. R. Lakshmanan and P. Nijkamp (eds), *Structure and Change in the Space Economy*, Berlin: Springer-Verlag, 155–73.

Chapman, G. P. 1977: *Human and Environmental Systems*. London: Academic Press.

Dearlove, J. 1979: *The Reorganization of British Local Government*. Cambridge: Cambridge University Press.

Dicken, P. 1999: *Global Shift*. London: Paul Chapman, 3rd edn.

Friedmann, J. 1986: The world city hypothesis. *Development and Change*, 17: 69–84.

Haggett, P. 1965: *Locational Analysis in Human Geography*. London: Arnold.

Honey, R. 1981: Alternative approaches to local government change. In A. D. Burnett and P. J. Taylor (eds), *Political Studies from Spatial Perspectives*, New York: Wiley, 245–74.

Lukermann, F. 1966: Empirical expressions of nodality and hierarchy in a circulation manifold. *East Lakes Geographer*, 2: 17–44.

Nystuen, J. D. and Dacey, M. F. 1961: A graph theory interpretation of nodal regions. *Papers and Proceedings of the Regional Science Association*, 7: 29–42.

Openshaw, S. and Taylor, P. J. 1979: A million or so correlation coefficients: Three experiments on the modifiable areal unit problem. In N. Wrigley (ed.), *Statistical Applications in the Spatial Sciences*, London: Pion, 120–38.

Sassen, S. 2000: *Cities in a World Economy*. Thousand Oaks, CA: Pine Forge Press.

Schattsschneider, E. E. 1960: *The Semi-Sovereign People*. Chicago: Dryden.

Storper, M. 1997: *The Regional World*. New York: Guilford Press.

Taylor, P. J. 1982: A materialist interpretation for political geography. *Transactions of the Institute of British Geographers*, n.s.7: 15–34.

Taylor, P. J. 1984: Introduction: political scale and political geography. In P. J. Taylor and J. W. House (eds), *Political Geography: Recent Advances and Future Directions*, London: Croom Helm, 1–7.

Taylor, P. J. 1994: The state as container: Territoriality in the modern world-system. *Progress in Human Geography*, 18: 151–62.

Taylor, P. J. 1995: Beyond containers: Inter-nationality, inter-stateness, interterritoriality. *Progress in Human Geography*, 19: 1–15.

Taylor, P. J. 1996: Embedded statism and the social sciences: Opening up to new spaces. *Environment and Planning A*, 28: 1917–28.

Taylor, P. J. 1997a: The crisis of boundaries: Towards a new heterodoxy in the social sciences. *Journal of Area Studies*, 11: 11–31.

Taylor, P. J. 1997b: Hierarchical tendencies amongst world cities: A global research proposal. *Cities*, 14: 323–32.

Taylor, P. J. 1999: "So-called world cities": the evidential structure within a literature. *Environment and Planning A*, 31: 1901–04.

Taylor, P. J. 2000a: World cities and territorial states under conditions of contemporary globalization. *Political Geography*, 19: 5–32.

Taylor, P. J. 2000b: Embedded statism and the social sciences 2: Geographies (and metageographies) in globalization. *Environment and Planning A*, 32: 1105–14.

Taylor, P. J. 2001a: Urban hinterworlds: Geographies of corporate service provision under conditions of contemporary globalization. *Geography*, 86: 51–60.

Taylor, P. J. 2001b: Specification of the world city network. *Geographical Analysis*, 33: 181–94.

Taylor, P. J. and Hoyler, M. 2000: The spatial order of European cities under conditions of contemporary globalization. *Tidjschrift voor Economische en Social Geografie*, 91: 176–89.

Taylor, P. J. and Walker, D. R. F. 2001: World cities: A first multivariate analysis of their service complexes. *Urban Studies*, 38: 23–47.

11 The Politics of Scale and Networks of Spatial Connectivity: Transnational Interurban Networks and the Rescaling of Political Governance in Europe

Helga Leitner

Today, virtually all modern nation-states and their subnational units have become increasingly enmeshed in larger patterns of global transformations and flows, affecting the nature of politics and governance and their geographies (Held et al., 1999). During the past decade, considerable research in Europe and the US has concerned itself with real and alleged transformations of the state and politics, and their changing geographies in the wake of economic globalization. Three issues in particular have been the focus of this work: the rise of supranational institutions, the declining power of the nation-state, and the growth of transnational networks. The second half of the twentieth century saw the proliferation of supranational institutions, such as the European Union (EU), ASEAN, United Nations, and the World Bank, resulting in new scales of political governance over and above that of the "traditional" nation-state. Economic globalization and the rise of supranational institutions also have been associated with a decrease in power of the nation-state relative to the global and subnational scales. Together, these developments suggest that the geographic scale at which political power and authority is located does not constitute a natural order, but rather is constructed and subject to change. Geographers have used the notion of the politics of scale to explain such sociospatial transformations.

Beyond this rescaling of political governance, however, we have also seen the rise of a new, horizontal mode of politics: transnational networks linking together state and/or civil institutions and actors across the boundaries that mark the traditional geography of political power (i.e., the bounded spaces of local, regional, national, and supranational state territories). Examples include transnational issue networks and social movements, such as human rights and environmental movements, as well as formalized transnational cooperative networks between cities and regions in Europe, linking cities and regions in and beyond Europe in complex ways. These transnational networks represent new modes of coordination and governance, a new politics of horizontal relations that also has a distinct spatiality. Whereas the spatiality of a politics of scale is associated with vertical relations among nested territorially defined political entities, by contrast, networks span space rather than covering it, transgressing the boundaries that separate and define these political entities.

In their analysis of sociospatial transformations, some geographers have subsumed such networks within the notion of a politics of scale. I suggest that this is problematic and agree with Brenner (2001: 592) that in the recent past we have seen an "analytical blunting, of the concept of geographic scale as it is applied, often rather indeterminately, to an expanding range of sociospatial phenomena, relations and processes." Subsuming distinct sociospatial projects, such as transnational interurban networking, the rescaling of power from the national to the supranational scale, claims to and defense of territory, all under the notion of a politics of scale, obscures the different spatialities implicated in these political geographic projects. In short, we need to conceptualize the politics of scale as only one dimension of a broader notion of spatial politics.

At the same time, however, we need to remember that while analytical distinctions can be made between different sociospatial projects, they should not be treated as independent from one another. To the contrary, I suggest that the focus of our analysis should be on examining the articulation of different sociospatial projects with one another. As I will go on to argue in this chapter, transnational networks among cities and regions have been shaped by power struggles among local, national, and supranational scales in Europe, but also are influencing these scalar politics.

In the following, I first discuss commonalties and differences in conceptualizations of the politics of scale in the geographic literature. I focus on differences in how power is conceptualized, where it is located, and the attention paid to concrete rhetorical and material practices and sociopolitical struggles in the (re)construction of scales. Second, I investigate selected aspects of network modes of governance, especially their spatiality and their articulation with the politics of scale. Throughout, I use the European Union as an example to develop these arguments, because this is a region where these processes are simultaneously unfolding with particular intensity.

Constructivist Perspectives on Scale

The conception of scale employed in research on the politics of scale departs dramatically from traditional notions that scales are simply different levels of analysis within which investigations of economic, social, and political processes are set. In this traditional view, geographic scale is a fixed, nested hierarchy of bounded spaces of differing size, such as the local, regional, national, supranational, and global (Leitner, 1997). Such a treatment assumes a stable and unchanging spatial structure of entities and permanencies. For example, in comparative politics, the national scale has been privileged, and policy variations among nation-states have been explained in terms of contextual differences at the national scale. This approach is problematic, because it ignores the relations among, as well as the influences of processes operating at, different scales (e.g., subnational and supranational), and how they interact to influence national policy.

The notion of geographic scale as a pregiven and fixed hierarchy of bounded spaces has come under increased scrutiny in human geography and has been replaced with a more dynamic conception, generally referred to as a constructivist perspective on scale (Delaney and Leitner, 1997). Although there remains a diversity of conceptions regarding its precise meaning, the common ground of this body of research is that:

- scale is socially constructed – not pregiven and fixed;
- the social and the spatial are mutually constitutive – for example the construction of a new scale of political governance, such as the European Union, involves the reconstruction of political relations among different scales of governance, and vice versa;
- the construction of scale incorporates power. The construction of scale, whether through material practices (production, distribution, consumption, regulation, surveillance, administration) or rhetorical practices (e.g., discourses on globalization or on the hollowing out of the nation-state) of individuals, groups and institutions, is a contested process that involves conflict-laden power struggles (Staeheli, 1994).

Scales are thus both the realm and outcome of social relations and struggles for control over social, political, economic, and geographic space (Swyngedouw, 1992). The concept of a *politics of scale* is used to connote that "geographical scales and scalar configurations are socially produced and politically contested through human social struggle..." (Brenner, 2001: 604). Central to the politics of scale is the manipulation of relations of power and authority by actors and institutions operating and situating themselves at different spatial scales. This process is highly contested, involving numerous negotiations and struggles between different actors as they attempt to reshape the spatiality of power and authority (Leitner, 1997).

The politics of scale involves both rhetorical and material practices. Rhetorical practices may employ scale as a way of framing/reframing reality. Scale framings such as the local or global are not merely rhetorical stances, however. They are neither neutral nor uncontested, and have real material consequences. For example, Jonas (1994) discusses the attempt by a multinational corporation headquartered in Worcester, Massachusetts, to frame itself as a local operation attached to the local community, in order to create support for its resistance to a takeover bid by a multinational conglomerate headquartered in Britain. By framing itself as a local operation, the corporation can be considered as engaging in a strategic representational practice, attempting to leverage the local to enhance its own power position vis á vis the foreign conglomerate (Delaney and Leitner, 1997).

In a second example, Andrew Herod's (1997) study of contract bargaining in the longshore industry focuses on the scalar strategies deployed by unions in labor contract negotiations.

> He describes the complicated emergence of a "national" scale of contractual coverage for the International Longshoreman Association out of prior port-by-port and regional agreements. The strategic issue concerned the scale at which bargaining takes place, but the ultimate objective was equalization of working conditions during an era of profound change in the nature (and politics) of shipping. Scale . . . becomes formalized in agreements over which issues should be covered by the master contract and which should be left to local bargaining units. The critical practices which produced this scalar division of responsibilities were negotiations backed up by direct actions of workers. (Delaney and Leitner, 1997: 96)

A final example of the politics of scale is the practices of state authorities during the 2000 spring meetings of the World Bank and the International Monetary Fund in Washington DC. In anticipation of large-scale protests by different social movements and networks, state authorities declared the space between the World Bank and IMF headquarters (across the street from each other) as international space, thereby justifying a military presence. Here we see a temporary redefinition or rescaling of a local space as a supranational space – a spatial strategy designed to contain the power of protestors.

Notwithstanding common reference in the geographic literature to the politics of scale there exist variations as to its precise meaning (Brenner, 2001; Marston and Smith, 2001), particularly with respect to how power is conceptualized, where it is located, and the degree of attention paid to concrete practices and struggles in the construction of scale. I will focus here on the latter two aspects.

A number of studies employing a constructivist perspective on scale conceptualize power as located exclusively in capitalist production relations (Harvey, 1982; Taylor, 1982; Smith and Dennis, 1987; Brenner, 1997; Brenner, 1998). Examining the historical geography of capitalism, Neil Brenner (1998) argues that different phases of capitalism are associated with distinct scale-configurations, or "scalar fixes." He identifies the following scale configurations reflecting different phases of capitalism. Between the late nineteenth century and the 1930s, the national scale was established as the most important geographical framework of capitalist territorial organization. From 1950 to the 1970s, the national scale continued to be the most important scalar fix for the Fordist-Keynesian round of capital growth. In contrast, since the 1970s we can observe a "denationalization of capitalist territorial organization and the pursuit of global scalar fixes" (Brenner, 1998: 473).

These scale configurations are seen as constituted and transformed by the sociospatial dynamic of capitalism. Thus at the heart of understanding scale production is an understanding of capitalist production. In this view, rescaling occurs when existing scalar fixes have become highly ineffectual for promoting the continued accumulation of capital – as during the world economic crises of 1873-96, the 1930s, and most recently, the 1970s (Brenner 1998: 473). Each of these major economic crises of the last century has significantly transformed the scalar organization of capitalism. Indeed, the rescaling that occurred can be viewed as an important geographic strategy of crisis displacement and crisis management on the part of capital and the state (Brenner, 1998: 477). At the center of such analyses, therefore, are observations about scalar changes in capitalist production relations, and their relationship to changes in the territorialization of social relations within and beyond nation-state boundaries.

Marston (2000) has been critical of this exclusive focus on capitalist production relations in theorizing rescaling and scalar reconfigurations. Instead, she proposes that relations of social reproduction and consumption also play a central role in the construction of scale, and that more attention needs to be paid to power relations outside the realm of capital-labor relations. While I agree that little attention has as yet been paid to social reproduction in research on the politics of scale, her latter assessment is questionable. In fact, in a large number of theoretical and empirical studies on the politics of scale, power is regarded as located in a range of actors and institutions multiply situated in economic, political and cultural contexts, with different stakes and ideologies (Agnew, 1993; Agnew, 1997; Herod,

1997; Jonas, 1994; Miller, 1994; Miller, 1997; Leitner, 1997; Smith, 1992; Swyngedouw; 1997; Brenner, 2001). While recognizing the situatedness of actors and institutions within capitalist political economic structures, these studies acknowledge the role of noneconomic structures, relations, and discourses in theorizing the politics of scale.

Similarly, studies analyzing the actual practice of scale construction, by actors attempting to effectuate and resist change, have not just focused upon the "power of capital, labor or the state – or some combination – as primary sites of scale construction" (Marston 2000: 221). For example, my analysis of the harmonization and rescaling of immigration policies from the national to the supranational scale in the EU details the role of transnational immigrant organizations and networks and nongovernmental organizations in the struggle over the rights of current and potential non-EU immigrants under a regime of harmonized immigration policies (Leitner, 1997). More generally, studies that pay attention to the actual practices of scale (re)construction by actors and institutions are less likely to reduce their theorizing of the politics of scale to power relations among capital, labor and the state. Of course, which power relations are being emphasized in theorizing the politics of scale will also depend on the sociospatial processes under investigation and the actors and institutions involved in attempting to effectuate change. In short, theorizing the politics of scale requires recognizing the dialectical relationship between structure and agency as played out in a variety of realms of society. The construction of the EU illustrates well how this operates.

The Construction of the European Union as a Politics of Scale

The history of European integration, and the emergence of the European Union (EU) as a new scale of political governance during the last forty years, is a paradigmatic example of the politics of scale. It demonstrates how scales of political governance are not fixed but are subject to change and are actively constructed.

The emergence of the EU has been the focus of many debates on economic globalization. It has been seen "... as a test case of the idea that scale incongruities between economic and regulatory processes can be tackled through creating larger scale state structures" (Low, 1997: 259). In other words, the EU can be seen as a supranational scale that has been developed in response to structural transformations of a capitalist economy that is global in scope and impact. Yet this economistic analytical lens on the construction of a supranational scale overlooks *inter alia* political motivations for the establishment of the EU – to ensure peace among European nation-states. It also conceals the political struggles over the making of the European Union that now have lasted more than forty years, and are still ongoing. In fact, the construction of this supranational scale (both rhetorically and materially), has been highly contested, involving numerous negotiations, tensions and struggles among different actors operating and situating themselves at different geographic scales (subnational, national, and supranational) as the following summary demonstrates.

At the center of these struggles, throughout, have been issues of whether, to what degree, and in which policy arenas, member states should surrender national sovereignty to centralized European Union institutions. Since the conception of the EU in the late 1950s, there has been a tug-of-war between EU institutions, particularly the European Commission, and member states. This has been a struggle over the geographic scale at which decisions about particular policies should be made (supranational or national), over the powers assumed by the institution entrusted with carrying out policies (Taylor, 1996), as well as over the principles according to which political power should be exercised (i.e., visions of justice and democratic accountability).

These struggles over the location of power and authority culminated, in the 1992 Treaty on European Union, in the formulation of the "principle of subsidiarity." The principle of subsidiarity defines the scales at which decisions should be made, and the powers invested in the different scales. This principle states that EU institutions are allowed to take action in areas that do not fall within their exclusive competence "only if and insofar as the objectives of the proposed action cannot be sufficiently achieved by the Member States and can therefore, by reason of the scale or effects of the proposed action, be better achieved by the Community" (Council of the European Union, 1997). Thus, in theory, the principle of subsidiarity affirms that power and authority should be located at the national scale, and can be transferred to the supranational scale only if it seems efficient and necessary. This was supposed to defuse fears among member states about the transfer of powers from the national to the supranational scale. In reality, however, the principle of subsidiarity is not only invoked by actors who oppose the transfer of power and authority to the supranational scale, but also has been used to justify increasing the competencies of EU institutions (Reichardt, 1994). It is clear, therefore, that the application and interpretation of the subsidiarity principle will continue to be contested, not only between EU institutions and member states but also among member states, and subnational scales of government (e.g., cities and regions). Recently, cities and regions in particular have attempted to make their voices heard through the "Committee of the Regions," thereby joining the struggles over the location and extent of power and authority within the EU (Tömmel, 1997).

Initial assessments of the outcome of these struggles and of the power balance between different scales have rendered contradictory conclusions. While some argue that member states have clawed back some of their autonomy (George, 1996), others contend that the member states did not succeed in decisively reining in the Commission (Marks, 1996). From my own research on EU regional policy and cohesion, it is clear that the European Commission remains one of the most active and influential players in this process of rescaling. Since the late 1980s, the Commission has created conditions and policy initiatives that not only increase its own power and authority, but also promote changes in power relations between national and subnational scales of government. For example, EU regional policy initiatives have attempted to strengthen the power of the local and regional scales vis á vis the national scale by providing resources to these scales, and by more directly involving subnational authorities in the EU policy process, specifically in the area of policy implementation. Significantly, these new EU institutional arrangements and policies can be thought of as disturbing and complicating traditional vertical relations between national and subnational units of governments and actors. One EU policy exemplifying this is the recent effort the EU has put into promoting transnational networks among cities and regions, to which I now turn.

EU-Sponsored Transnational Networks between Cities and Regions

There currently exists an influential public policy discourse in Europe, in which policy networks are being widely presented as a solution to social, economic, and political governance problems. This discourse draws selectively on academic research on economic and policy networks, heralding the benefits of network modes of governance. In this literature, networks are variously described as more efficient, more flexible, and more effective means of assembling resources and actors to complete complex tasks, than economic competition (markets) or centrally directed structures (hierarchies). They are said to promote and involve collective rather than individual action, and horizontal, negotiated self-organization, characterized by consensual and harmonious relations. German social scientists, in particular, have argued that, in a world characterized by increasing interdependence and dissolution of the distinction between the state and civil society, policy networks are a superior mode of governance, better suited to generating economic growth and resolving social problems than traditional hierarchical relations of governance (Martin and Mayntz, 1991; Mayntz, 1994).

This highly optimistic view about the benefits of economic and policy networks, and the positive attributes conferred upon them, has recently come under criticism, and has been tempered by a more agnostic view (Rhodes and Marsh, 1992; Marsh, 1998). Marsh and Rhodes (1992) argue convincingly that networks with the above characteristics in fact constitute ideal types, against which actually existing network modes of governance must be examined. Similarly, alleged network benefits must be ascertained empirically. Other scholars have been critical of the tendency to present networks as distinctly different from hierarchies and markets, pointing out that markets, hierarchies and networks, while different modes of governance, "do not exist in isolation, but are necessarily articulated" (Hay, 1998: 39; see also Leitner, Pavlik, and Sheppard, 2002).

Notwithstanding these reservations, the network ideal still is presented as the norm in public policy discourse, where the focus is on the benefits of networks and their efficacy for addressing problems in social, economic, and political governance. In the EU, as indicated above, networking and networks are not only fashionable concepts in public policy discourse, but also have become a key part of the policy process. Networks of coordination and governance among firms and between the public and private sectors have been implemented at various geographic scales, from the local to the transnational, including interurban and interregional networks. Since the mid 1980s, the European Union's Directorate General for Regional Policy and Cohesion (DG XVI) alone has introduced four major interurban and interregional network programs, providing funding[1] to a total of several thousand transnational network projects with limited time horizons: ECOS and OUVERTURE, PACTE, RECITE, and INTERREG.

ECOS and OUVERTURE (external interregional cooperation)

These programs aim to establish networks between regions and cities in the EU and Central and Eastern Europe in order to facilitate exchange of information on a wide range of topics. Approximately 250 projects have been funded, involving more than 1000 regions and cities, and addressing a wide range of themes including local and regional services, local economic development, sectoral economic development, tourism strategies, environment and energy (European Commission, 1994b).

PACTE (exchange of experience)

This program has funded almost 400 cooperative networks among regional and local authorities within the EU since its inception in 1990. Initially its main objective was to promote exchange of know-how and experience among network partners on a wide range of topics, including economic development, environment, energy and natural resources, health and social policy, urban planning, public administration, transport and technology, and research. Once initial exchanges had been undertaken and projects completed, it was envisaged that partners would then move on to join other programs such as a RECITE network (European Commission, DGXVI, 1996).

RECITE (regions and cities for Europe)

This program includes approximately 37 different network initiatives between cities and regions of various sizes (but a minimum of 50,000 people) in different member states. Its urban networks address issues and engage in actions designed to foster economic development, to find solutions to housing and unemployment problems, and to more generally lobby the EU, member states and private businesses in order to promote the common interests of network members (European Commission, DG XVI, 1997).

INTERREG (interregional and cross-border cooperation)

The key aim of the INTERREG Community Initiatives (I and II) has been to develop cross-border cooperation in order to help regions located on the EU's internal and external frontiers overcome specific problems arising from their relative isolation. INTERREG projects address a wide range of issues, including technology transfer, the development of small and medium-sized enterprises, development of transnational land administration systems and regulations as a basis for land-use, planning economic development and land management, and controlling flooding and combating drought (European Commission, 1994a).

These are just four transnational network programs supported by one Directorate. Various other EU Directorates also sponsor transnational urban and regional networks. For example, the Directorate General for

External Relations (DG I) supports networking among towns and regions in the Mediterranean area through its Med-Urbs program, and has also been preparing to launch programs including local authorities and islands as far away as Asia and Latin America.

Although cooperation among network members occurs around a wide range of issues, we can distinguish two major categories of networks. *Thematic networks* link together places with common concerns and problems, irrespective of their location. One example is the Demilitarized network – involving sixteen different cities and regions in five member states, all affected by the restructuring of defense-related industries and the closure of military bases. Another example is the Quartiers en Crise (Neighborhoods in Crisis) network that brings together 30 cities concerned with developing integrated approaches to combating housing and unemployment problems and social exclusion in urban neighborhoods. *Territorial networks* link together places in a common geographic region or in particular types of regions. For example, the cross-border networks of the INTERREG programs normally extend over at least two countries (see Figure 11.1), and the Atlantic Area encompasses 41 regions in France, Spain, Portugal, UK, and Ireland along the Atlantic coast (see Figure 11.2).

A major distinction between these two types of networks is their geographic scope, which varies from spatially contiguous clusters of cities and regions, to networks of pan-European scope, some extending beyond the EU, and networks of global scope stretching to cities in Latin America and Asia. This suggests that networks are themselves scaled, from the local to the global. Network scales are emergent properties of sociospatial processes operating inside and beyond networks (Leitner, Pavlik, and Sheppard, 2002). For example, in the case of the interurban networks discussed here, network scales are to a large degree constructed by directives of the European Commission.

As noted above, network activities range from the exchange of information and expertise (the most common network activity, typified by ECOS, OUVERTURE, and PACTE) to collaborative projects (INTERREG, RECITE). These differences in network activities and general agendas are shaped to a large extent by the EU directives under which the networks were set up. Thus ECOS, OUVERTURE and PACTE were originally intended to primarily foster the exchange of information, whereas RECITE and INTERREG were designed not only to begin cooperation, but also to lead to joint projects with tangible results.

While mostly stimulated at the initiative of the EU, these networks also function as strategic alliances among cities and regions, forged around a common agenda perceived to be of mutual advantage. In terms of political space they constitute a new, horizontal layer of politics for urban and

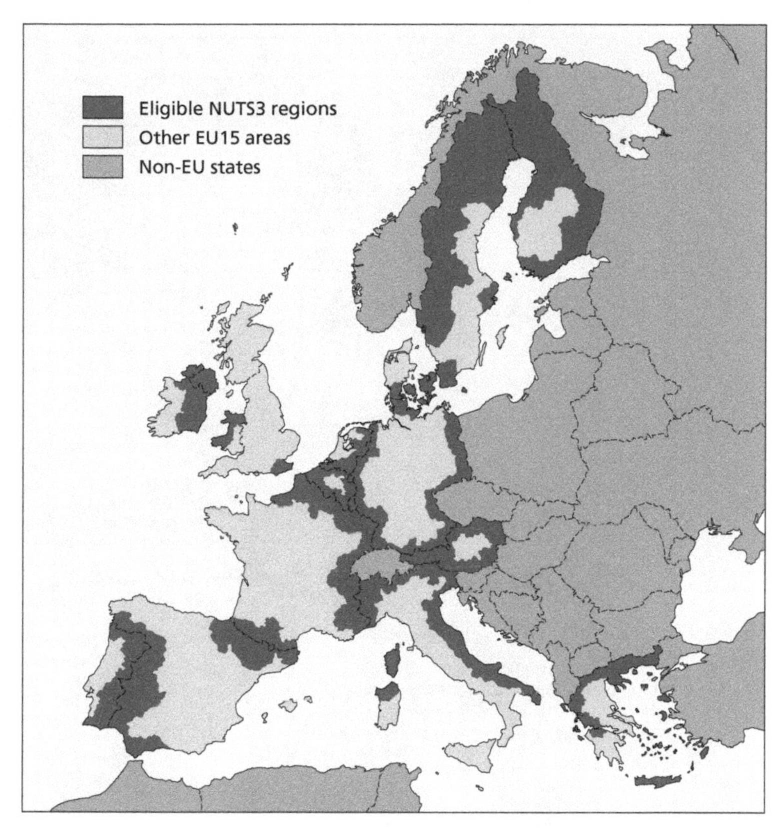

Figure 11.1 *Territorial networks INTERREG IIIA cross-border regions.*
Source: http://europa.eu.int/comm/regional_policy/interreg3/carte/cartes_en.htm

regional authorities, in addition to the traditional politics of vertical (hierarchical) relations between different tiers of the state.

Network Spaces: Transgressing and Challenging Boundaries

The geography of the newly emerging complex political spaces associated with networks defies easy description, but does imply a different geography than that of the familiar political map. This map organizes and divides the world into territorial states – with distinct boundaries enclosing spatially contiguous national territories. The territorial state itself is a multiscalar mode of political governance, characterized by nested layers of contiguous

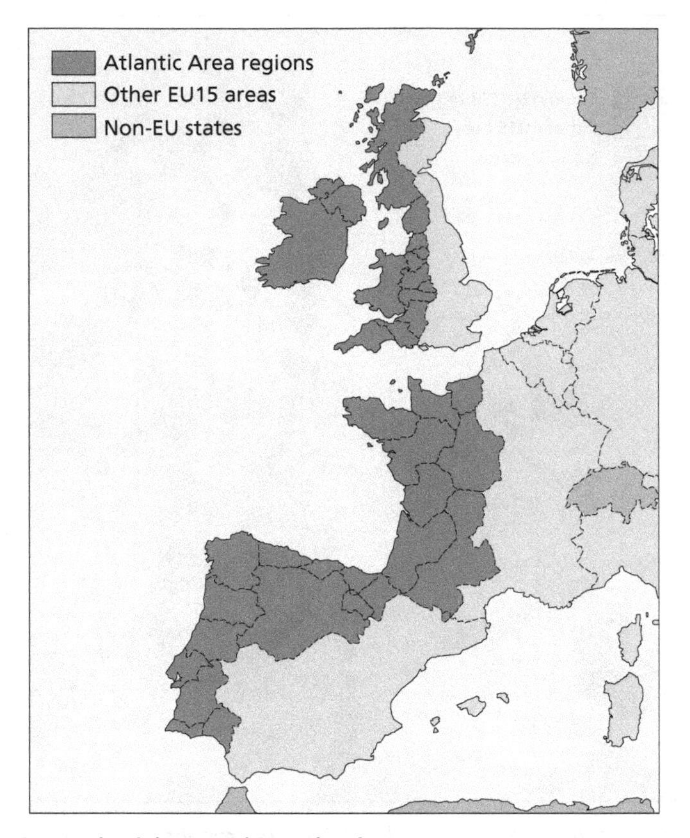

Figure 11.2 *Territorial networks – Atlantic Area.*
Source: http://europa.eu.int/comm/regional_policy/interreg3/carte/cartes_en.htm

bounded spaces with different degrees of permeability – the local, regional, and national tiers of the state. In contrast, network modes of governance represent a relational space, transcending the local, regional, and national boundaries of the territorial state. Thus, networks should be thought of as *spanning space* rather than covering it. While membership in a network can be defined, it generally cannot be mapped as a bounded territory (except for a spatially contiguous network), but must be represented as a disjoint set of local territorial units connected to one another (see Figure 11.3). As the case of transnational EU networks among cities and regions demonstrates, the spatial surface spanned by networks is also fluid and unstable, as different members join and leave each network and networks are born and terminated. Members of any interurban network can also belong to many other networks, meaning that networks overlap and intermingle with one

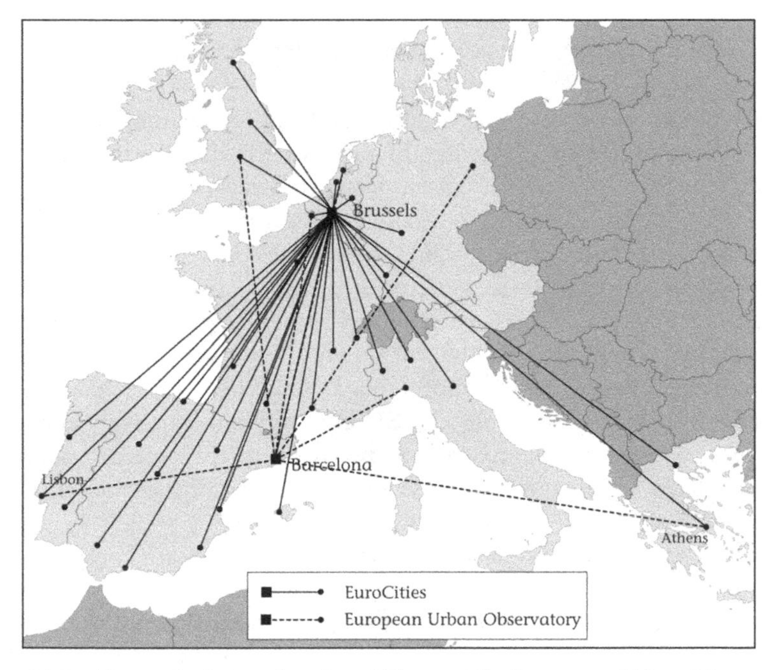

Figure 11.3 *Thematic Networks – EuroCities and the European Urban Observatory.*
Source: European Commission, DG XVI (1997)

another in complex ways. For example, Brussels, Barcelona, Lisbon and Athens are involved in multiple networks. Figure 11.3 shows the complex spaces of connectivity of two such networks, EuroCities[2] and the European Observatory.[3] This is quite different from the familiar political map, where different territories at any one scale are independent units that only belong to a single territorial unit at each higher scale. The outcome is a complex maze of transnational networks between cities and regions that defies easy description and delimitation in terms of its geography (Leitner, Pavlik, and Sheppard, 2002).

Most transnational interurban networks are still in their early stages of development and may be considered an embryonic form of governance, making it too early to draw definite conclusions regarding their implications for network participants, or for reconfigurations of political space and scalar relations. Notwithstanding this proviso, as I will show below, transnational networks between cities and regions in Europe are creating new geographies of governance that are articulated in complex ways with scalar (vertical) modes of governance.

Networks and the Politics of Scale

Transnational networks between cities and regions in the EU cannot be seen as separate from the scalar hierarchies of the EU political space. Rather, networks are dialectically related to scalar structures. Networks and scalar configurations are mutually constitutive of one another in complex ways. Networks help construct and contest scales and (re)configure scalar relations. In turn scalar structures construct and contest networks. In the case discussed here, interurban networks may help strengthen the power of the local (urban and regional) scale vis á vis the national and supranational scales, thereby contributing to a reconfiguration of scalar relations. In turn the supranational scale, the EU commission, is actively engaged in constructing transnational interurban networks to enhance its own power, authority, and legitimacy relative to other scales, and member states also contest such scalar reconfigurations.

Once in existence, transnational networks present an opportunity for cities and regions to strengthen their power and authority vis á vis the national government and the European Commission. Traditionally, particularly in highly centralized European states such as the UK, local and regional authorities have experienced little autonomy from the central state. Recent studies suggest that EU-sponsored transnational networks have begun to change this, both strengthening the power of local authorities vis-à-vis the national scale, at least in certain policy arenas, and also increasing their capacity to directly influence the supranational scale, i.e., EU Commission policy making. In a study of three cross-border networks between French and British local authorities supported by the INTERREG program, Church and Reid (1996: 1310) find that British local officials believe that transnational networks do strengthen their political autonomy vis á vis central government, especially in the area of local economic policy. Benington and Harvey (1998) suggest that the collective efforts of transnational networks also have played an important role in the formulation of some EU Community Initiatives. For example, the Demilitarised network played an important role in the formulation of KONVER II, a Community Initiative providing EU funding for regions attempting to counter the effects of a declining defense industry, to promote industrial conversion, and to find ways to reuse military facilities. Transnational urban and regional networks not only empower participating cities and regions relative to other scales, but also empower them by comparison to cities and regions left out of networks. They may also strengthen the power of localities vis-à-vis mobile capital, helping overcome the inability of local entrepreneurialism to realize local prosperity in the face of deregulated financial spaces and mobile capital (Leitner and Sheppard, 1999).

The relationship between these networks and hierarchical scales of governance works in both directions, however, because most transnational interurban and regional networks owe their existence to the European Commission. The Commission not only helps create and financially supports networks, but also uses its regulatory powers to closely circumscribe and monitor their goals, network agendas, types of cooperation, and geographic reach. This has the effect of promoting policies and policy changes that further Commission goals and institutional interests. The tendency of "bias" in EU policies towards strengthening the EU as a whole has been observed for other policy arenas (Dang-Nguyen et al., 1993). The direct involvement of the Commission with cities and regions potentially also increases its legitimacy in the eyes of such localities, and helps foster greater identification of cities and regions with the EU. Hence the promotion of transnational networks by the EU can also be seen as a strategy to stretch the political identity of local authorities beyond their traditional loyalty to the nation, to embrace a supranational, European, identity. This strategy parallels the development of national identity during processes of nation building, but at the supranational scale (Hobsbawm, 1994; Smith, 1991).

For their part, member states also attempt to contain and contest the increasing influence of both EU and subnational institutions as a result of EU-sponsored transnational networks. They are concerned with both how such networks empower local and regional authorities relative to the national state, and with any Europeanization of local political identities. Indeed, some national governments, such as the British government, have blocked network initiatives, and thus the direct access of cities and regions to EU funding (Benington and Harvey, 1998).

Conclusion

In this chapter, I have sought to demonstrate the utility and limitations of the concept of a politics of scale for understanding and explaining sociospatial transformations, specifically transformations in political governance, using the European Union as a case study. Geographers have used the politics of scale to understand a variety of such sociospatial transformations, including shifting global-local relations, the rise of supranational organizations and regulation, the reterritorialization of labor regulation, and the shifting importance of the nation-state. In the process, much effort has been devoted to theorizing scale construction, scalar transformations, and rescaling processes. Some studies have focused primarily on structural forces underlying scale construction and scale transformations, with causal powers primarily located in capitalist production relations. The majority of studies conceptualizing scale construction has emphasized the structuration of scale production, however. While recognizing the embeddedness of agency within political and economic structures and relations, attention is also paid to noneconomic relations and the actual practices of actors and institutions involved in scale constructions. The latter is particularly important since we should not lose sight of how actual practices construct the broader structures that are the focus of grand theorizing. I use the construction of the supranational scale of the European Union as an example to demonstrate the importance of a structurationist approach to scale construction and scalar reconfigurations in political governance.

Transformations in political governance, however, have multiple geographies. Alongside scalar reconfigurations in political governance, networks of spatial connectivity have assumed increasing importance as forms of political governance and resistance. By comparison to the familiar scaled political map, such networks have a different spatiality, spanning space rather than covering it. They connect places horizontally across the bounded spaces of political territorial entities, which themselves are part of scalar state structures. Networks of spatial connectivity thus constitute a distinct sociospatial project that cannot be subsumed under scale politics.

Yet networks also cannot be separated from other sociospatial projects such as scalar reconfigurations. In fact, networks and scales are closely articulated with one another. As I have shown in the last section of the chapter, transnational networks among cities and regions and scalar state structures mutually constitute and affect one another in complex ways in the European Union. Transnational interurban networks help empower local authorities by engaging them in new spatial collaborations that challenge the scalar power relations of EU member states, and even the EU itself. At the same time, these networks are embedded in a politics of scale – struggles over political power between their sponsor (the European Commission), member states, and local authorities. This process of mutual constitution will continue to be fraught with tensions and struggles between different actors and institutions, as they attempt to reshape the spatiality of power and authority. Recent developments in the European Union thus exemplify well that spatial politics entails a variety of distinct sociospatial projects that are bound up with, but not reducible to, a politics of scale.

Notes

1 Networks are allocated funding from within the EU programs on a competitive basis.

2 The EuroCities network was formed to represent the interests of "second tier" cities, i.e., large and mainly noncapital cities, within the EU. It has a membership of over 40 cities (including Barcelona, Madrid, Amsterdam, Rotterdam, Birmingham, Edinburgh, Bologna, Milan, Lisbon, Strasbourg, Montpellier, Frankfurt, Athens, Thessaloniki) and has engaged in lobbying various EU Directorates and promoted joint concerns at well publicized conferences. In contrast to most other transnational urban networks, EuroCities started as a result of a bottom-up initiative of the city of Rotterdam in 1986.

3 European Urban Observatory links cities for the purpose of the development of a shared database and system to aid decision-making (including Barcelona, Genoa, Berlin, Athens, Amsterdam, Lille, Lisbon, Birmingham, and Brussels).

References

Agnew, J. 1993: Representing space: Space, scale and culture in social science. In J. Duncan and D. Ley (eds), *Place/Culture/Representation*, London: Routledge, 251–71.

Agnew, J. 1997: The dramaturgy of horizons: Geographical scale in the "reconstruction of Italy" by the new Italian political parties, 1992–1995. *Political Geography*, 16: 99–121.

Benington, J. and Harvey, J. 1998: Transnational local authority networking within the European Union: Passing fashion or new paradigm? In D. Marsh (ed.), *Comparing Policy Networks*, Buckingham/Philadelphia: Open University Press, 149–66.

Brenner, N. 1997: State territorial restructuring and the production of spatial scale: Urban and regional planning in the FRG 1960–1990. *Political Geography*, 16: 273–306.

Brenner, N. 1998: Between fixity and motion: Accumulation, territorial organization and the historical geography of spatial scales. *Environment and Planning D: Society and Space*, 16: 459–81.

Brenner, N. 2001: The limits to scale? Methodological reflections on scalar structuration. *Progress in Human Geography*, 25: 591–614.

Council of the European Union, 1997. Article 5(ex Article 3b): The principle of subsidiarity. In the *Consolidated Version of the Treaty Establishing the European Community*. Brussels: General Secretariat of the Council, 1999. (http://ue.eu.int/en/treaties/htm)

Church, A. and Reid, P. 1996: Urban power, international networks and competition: The example of cross-border cooperation. *Urban Studies*, 33: 1297–1318.

Dang-Nguyen, G., Schneider, V. and Werle, R. 1993: Networks in European policy making: Europeification of telecommunications policy. In S. S. Andersen and K. A. Eliassen (eds), *Making Policy in Europe: The Europeification of National Policy-making*, London: Sage Publications, 93–114.

Delaney, D. and Leitner, H. 1997: The political construction of scale. *Political Geography*, 16: 93–7.

European Commission 1994a. *Interregional and Cross-Border Cooperation in Europe*. Luxembourg: Office for Official Publications of the European Communities.

European Commission 1994b: *Review of Interregional Cooperation: ECOS – Ouverture*. Conference Proceedings, July. Luxembourg: CEC.

European Commission, DG XVI 1996. *Review of Interregional Cooperation: PACTE Program*. Conference Proceedings, Review # 7. Luxembourg: CEC.

European Commission, DG XVI 1997: *Review of Interregional Cooperation. Achievements under RECITE I*. Brussels: ECOTEC Research and Consulting Ltd.

George, S. 1996: *Politics and Policy in the European Union*. New York: Oxford University Press.

Harvey, D. 1982: *The Limits to Capital*. Chicago: University of Chicago Press.

Hay, C. 1998: The tangled webs we weave: The discourse, strategy and practice of networking. In D. Marsh (ed.), *Comparing Policy Networks*, Buckingham/ Philadelphia: Open University Press, 33–51.

Held, D., A. McGrew, D. Goldblatt, J. Perraton 1999: *Global Transformations: Politics, Economics and Culture*. Stanford, CA: Stanford University Press.

Herod, A. 1997: Labor's spatial praxis and the geography of contract bargaining in the US east coast longshore industry. *Political Geography*, 16: 145–69.

Hobsbawm, E. J. 1994: The nation as invented tradition. In J. Hutchinson and A. D. Smith (eds), *Nationalism*, Oxford and New York: Oxford University Press, 76–83.

Jonas, A. 1994: Editorial: The scale politics of spatiality. *Environment and Planning D: Society and Space*, 12: 257–64.

Leitner, H. 1997: Reconfiguring the spatiality of power – The construction of a supranational migration framework for the European Union. *Political Geography*, 16: 123–43.

Leitner, H. and Sheppard, E. 1999: Transcending interurban competition: Conceptual issues, and policy alternatives in the European Union. In D. Wilson and A. Jonas (eds), *The Urban Growth Machine – Critical Perspectives Two Decades Later*, Albany, NY: State University of New York Press, 227–46.

Leitner, H., Pavlik, E. and Sheppard, E. 2002: Networks, governance and the politics of scale: Interurban networks and the European Union. In A. Herod and M. Wright (eds), *Geographies of Power: Placing Scale*. Oxford: Blackwell, 274–303.

Low, M. 1997: Representation unbound: Globalization and democracy. In K. Cox (ed.), *Spaces of Globalization: Reasserting the power of the local*, London: Guilford, 240–80.

Marks, G. 1996: Exploring and explaining variation in EU cohesion policy. In L. Hooghe (ed.), *Cohesion Policy and European Integration: Building Multi-Level Governance*, New York: Oxford University Press, 388–422.

Marsh, D. 1998: The development of the policy network approach. In D. Marsh (ed.), *Comparing Policy Networks*, Buckingham/Philadelphia: Open University Press, 3–20.

Marsh, D. and Rhodes, R. A. W. 1992: *Policy Networks in British Government.* Oxford: Clarendon.

Marston, S. 2000: The social construction of scale. *Progress in Human Geography*, 24: 219–42.

Marston, S. and Smith, N. 2001: State, scales and households: limits to scale thinking? A response to Brenner. *Progress in Human Geography*, 25: 615–19.

Martin, B. and Mayntz, R. (eds) 1991: *Policy Networks: Empirical Evidence and Theoretical Considerations.* Frankfurt: Campus.

Mayntz, R. 1994: *Modernization and the Logic of Interorganizational Networks.* MIPFG Working Paper No. 4. Köln: Max-Planck Institute für Gesellschafts-forschung.

Miller, B. 1994: Political empowerment, local-central state relations, and geographically shifting political opportunity structures: Strategies of the Cambridge, Massachusetts peace movement. *Political Geography*, 13: 393–406.

Miller, B. 1997: Political action and the geography of defense investment: Geographical scale and the representation of the Massachusetts Miracle. *Political Geography*, 16: 171–85.

Reichardt, W. 1994: Die Karriere des Subsidiaritätsprinzips in der Europäischen Gemeinschaft. *Österreichische Zeitschrift für Politikwissenschaft*, 1: 53–66.

Rhodes, R. A. W. and Marsh, D. 1992: New directions in the study of policy networks. *European Journal of Political Research*, 21: 181–205.

Smith, A. D. 1991: *National Identity.* Harmondsworth: Penguin.

Smith, N. and Dennis, W. 1987: The restructuring of geographical scale: Coalescence and fragmentation of the northern core region. *Economic Geography*, 63: 160–82.

Smith, N. 1992: Geography, difference and the politics of scale. In J. Doherty, E. Graham, and M. Malek (eds), *Postmodernism and the Social Sciences*, New York: St. Martin's Press: 57–79.

Staeheli, L. 1994: Empowering political struggle: Spaces and scales of resistance. *Political Geography*, 13: 387–91.

Swyngedouw, E. 1992: The mammon quest: "Glocalisation," interspatial competition and the monetary order: the construction of new scales. In M. Dunford and G. Kafkalas (eds), *Cities and Regions in the New Europe*, London: Belhaven Press, 39–68.

Swyngedouw, E. 1997: Neither global nor local: "Glocalisation" and the politics of scale. In K. Cox (ed.), *Spaces of Globalization: Reasserting the power of the local*, London: Guilford, 137–66.

Taylor, P. J. 1982: A materialist framework for political geography. *Transactions of the Institute of British Geographers*, n.s.7: 15–34.

Taylor, P. 1996: *The European Union in the 1990s.* New York: Oxford University Press.

Tömmel, I. 1997: The EU and the regions: Towards a three-tier system or new modes of regulation? *Environment and Planning C: Government and Policy*, 15: 413–36.

12 Scale and Geographic Inquiry: Contrasts, Intersections, and Boundaries

Eric Sheppard and Robert B. McMaster

> In that Empire, the craft of Cartography attained such Perfection that the map of a Single Province covered the space of an entire City, and the Map of the Empire itself an entire Province. In the course of Time, these Extensive maps were found somehow wanting, and so the College of Cartographers evolved a Map of the Empire that was of the same Scale as the Empire and that coincided with it point for point. Less attentive to the Study of Cartography, succeeding Generations came to judge a map of such Magnitude cumbersome, and, not without Irreverence, they abandoned it to the Rigours of sun and Rain. In the western Deserts, tattered Fragments of the Map are still to be found, Sheltering an occasional Beast or beggar; in the whole Nation, no other relic is left of the Discipline of Geography.
>
> (Jorge Luis Borges, "Of exactitude in science")

As the chapters of this book demonstrate, conceptions of geographic scale range across a spectrum of almost intimidating diversity. While sharing the notion that scale matters, and ranging from geographically localized phenomena to the globe, little else seems to remain the same as discussion shifts from the representative fraction to the construction of scale. (Indeed, as discussed in the introduction, even the definition of large and small scale is reversed!) This reflects the diversity of geographic inquiry in general. Geographers have long worried, for example, about whether human and physical geographers have much in common and about the place of cartography within geography, and nongeographers frequently puzzle about, and challenge geographers to articulate, the nature (and boundaries) of their discipline. Is Geography natural science, social science, or cultural studies, and what makes it distinct from other disciplines?

There is no question that different traditions of thinking about scale have emerged within different communities of geographers, reflecting divergent traditions of theory, philosophy, epistemology, and methodology. To some degree these also can be associated with different subdisciplines, thereby reinforcing the notion that Geography is a diverse or even fractionated field. Yet, as we attempt to indicate here, such boundaries and distinctions do not hold up well to closer analysis: there is, in fact, considerable overlap within this diversity. In the following we summarize these differences, point to some common ground, and speculate on the boundaries of scale as an element of geographic inquiry.

Contrasting Approaches to Scale

Geographic information science and cartography

Within the literature on cartography and geographic information science, discussions of scale focus on questions of definition, measurement, representation, and analysis. As Lam and Goodchild both discuss, scale is used to define how a map relates to the earth's surface (e.g., the representative fraction and the *S/L* ratio), the scale or extent of a study area (*L*), and the resolution of the finest spatial data available (*S*). It then becomes possible to measure the amount of information in a cartographic image, the complexity of the patterns on such images (e.g., using fractals), and the smallest resolution at which reliable spatial analysis is possible. The question of representation, how a map will deviate from the complexity of reality, is tackled in part by deciding which features, at which levels of detail, should be present on maps of different cartographic scales – the map generalization problem. (Of course, maps as representations entail many other issues also, cf. Monmonier, 1991; Wood, 1992) Researchers are far from being able to identify the appropriate information content for various scales, given a specific map purpose and audience. Furthermore, the Modifiable Areal Unit Problem (discussed by Taylor and in the Introduction) has stimulated an extensive body of research into how to incorporate scale into methods of spatial analysis, investigating such questions as:

- can we identify the operational scale of a geographic phenomenon, and how is this determined? (Is it, for example, the scale at which this phenomenon shows the most spatial variation?)
- how do relationships between variables change as the scale of measurement (resolution) increases or decreases?

- to what degree can information on spatial relationships at one scale be used to make inferences about relationships at other scales?

Within this literature, analysts commonly use two different geographic data models: the vector model that creates a map-like representation of points, lines, and areas, and the raster model that "tesselates" the surface of the earth into similar units. For instance, units at the same geographic scale are conceived as regular polygons covering the map (e.g., a raster system). The size of these units is fixed, and measured by Euclidean distance (or occasionally, spherical coordinates). Larger scale units typically have the same shape as smaller scale units, and are derived by aggregating together a constant number of smaller scale units (see Walsh et al., this volume). Scale is thus conceived of as a hierarchy of levels, in which the size of units at each scale is fixed, the shape of such units is geometrically regular and identical (aside from certain specific data models, such as the quadtree), and smaller scale units nest within larger scale units.

Physical geography: hierarchy theory and spatial analysis

In physical geography, and the spatial analytic tradition of human geography, these ideas are influential, but undergo modification because of the need to understand substantive geographic processes. Of the four scale issues in the earth sciences discussed in the introduction to this volume, the chapters here each focus on issues of multiscale analysis. Phillips summarizes conditions under which different scales can in fact be modeled separately, but also stresses the need for a multiscalar perspective in order to interpret model results adequately. Walsh et al. show that spatial correlations can vary across scale when different independent variables have different operational scales, and Easterling and Polsky propose a multilevel statistical model to examine the joint effect of independent variables operating at different scales. Finally, Zeigler et.al. discuss the concept of fundamental scale elements (such as observation grain and extent, attribute type grain) and how these assist us in understanding the embedded scales on biogeographical maps.

Physical geographers typically seek to model spatial processes, and thus must pay attention to temporal as well as spatial scale. In addition, their concern for particular geographic processes requires attention to what constitutes meaningful units of analysis for particular phenomena. Thus hydrologists concern themselves with watersheds of various scales, ecologists with ecosystems, landscape analysts with landscape complexes, and climatologists with climate or vegetation regions. In each case, it is possible to conceive of a hierarchy of, say, watersheds at various scales, from that of a minor tributary to that of the Amazon. As in the case of GISc, watersheds at one scale completely cover the landscape, and are nested within higher scale watersheds. These are not regular geometric polygons, however, but take a variety of shapes that reflect the biophysical and social processes shaping nature. Also, as noted by Zeigler et. al., the data with the greatest uncertainty establish the uncertainty of the entire process.

Three of these chapters invoke hierarchy theory as a guiding principle of analysis. As discussed in the Introduction, hierarchy theory is commonly invoked across the earth sciences as a way of describing nature, of particular utility for scale-based analysis. To recall, hierarchy theory enables natural phenomena to separate into distinctive time and space scales, and implies that different processes are expected to have characteristic operational spatiotemporal scales at which they operate. This means that multiscalar analysis can be dramatically simplified. It has been even suggested that nature can typically be approximated by hierarchy theory because it is an evolutionarily successful strategy. H. A. Simon (1962) illustrates this claim with a parable of two watchmakers. He argues that a watchmaker who tries to build an entire watch at once will be much less successful than one who adopts a hierarchical approach; building subcomponents that can be set aside and subsequently assembled together into larger and larger subunits. He terms the latter approach as more robust, since if this watchmaker is interrupted in the middle (s)he will not have to begin again from scratch.

Some physical geographers do not agree, however, that hierarchy theory provides a universal scientific approach to analyzing nature. Michael Church (1996) argues that different space/time scales should be analyzed in different ways. Noting that "our construction of rational order in the world around us is essentially constrained by the scales of space and time within which we examine the world" (p. 167), and pointing also to limitations in our ability to observe both very small/short space-time scales and very large/long scales, he proposes four distinctive "modes of theory construction":

> At small space and time scales, phenomena are recorded in sequences which describe very large numbers of characteristic events. Descriptions are statistical, and processes are considered to be stochastic. At the scales of classical mechanics, deterministic theories are sustainable. At still larger scales, system evolution reveals contingent endogenous effects which cannot be predicted, even though the system remains deterministic. Nonlinear dynamical models, expressing chaotic behaviour, are appropriate. At the largest scales of space and time, landscape evolution is entirely contingent, and we adopt a narrative,

> particularistic model of explanation. Each level of theory construction must be consistent with the others if the subject is to present a viable construction of nature, but it is not obvious that phenomena described at each scale can be derived from theory at different scales. (p. 147)

In short, instead of seeking to simplify medium-number complex nonlinear systems to small-number systems at all scales of analysis, Church suggests that appropriate models should shift from large-number systems, to small-number systems, medium-number systems and historical narratives, as we move from the bottom to the top of the hierarchy. Phillips (this volume) also suggests that interpretation is a vital step, and that this requires a multiscale perspective.

To summarize, in this second tradition of physical geographic research on scale, the complexity of nature is often tackled by adopting the proposition of hierarchy theory that nature can be divided into a hierarchy of space-time scales of progressively increasing extent. Here, again, scale is a hierarchy of levels. Lower levels need not nest within higher levels (Wu, 1999), and the units at each scale are natural regions rather than geometric polygons, but in practice they are treated as fixed and very often as nested. This also adopts a reductionist view of causality, typical of the natural sciences, a form of methodological individualism in which "the higher dimensional object is somehow 'composed' of... lower dimensional [phenomena], which have ontological primacy" (Levins and Lewontin, 1985: 271). In hierarchy theory, causality runs from the bottom up. Smaller scale (more micro) phenomena and processes "provide the initiating conditions" (Wu, 1999: 271) for larger scale phenomena.

Human geography: constructionist approaches

Some of the influential early research on scale and spatial analysis was initiated by human geographers (Taylor, this volume), but in recent years many human geographers have turned away from attempting to construct elegant value-free explanatory models of reality. In Church's (1996) terms, their theoretical strategy has become that of the historical narrative rather than the recursive causal model. Rather than seeking to reduce the complexity of the world in order to make it amenable to analytical modeling, they seek to grasp that complexity. Rather than beginning with a fixed set of categories and objects with given properties and then seeking to divine the causal connections between them, they question how such phenomena and properties come into existence, and how their significance and properties change over time as a result of their relations to other phenomena. It is important to point out that such questions are not only posed in the human sciences: Levins and Lewontin (1985) and Smolin (1997), for example, insist that these are also central questions in biology and physics. Yet while such voices are rare in the physical sciences, they have become common in many human sciences.

The common feature of the research reported by human geographers in these chapters is to raise these questions about scale. A contructionist approach to scale seeks to ask how particular scales come into existence for particular phenomena, and how their relative importance changes over time. Swyngedouw (this volume) provides examples of how this applies to the physical environment; where the scales at which water is managed, and indeed the shape of watersheds themselves, change as a result of political and economic processes. Goodchild argues that even in our digital age, when distance no longer is an impediment to the movement of information, in some ways local information, and thus the local scale, is becoming more important. For example, many now are forecasting that the widespread use of mobile phones and personal digital assisants (PDA) means that firms can attract local customers as they walk down the street by sending messages to their mobile devices, turning shopping into a more local experience again, and away from large-scale retailers such as Amazon.com or Ebay (Goodchild, 2000). Such trends and struggles shape the scales, in this case, at which consumption occurs. Marston, Smith, Taylor, and Leitner show, respectively, how processes of social reproduction, capital accumulation, corporate strategies, and political struggle have shaped and are shaping the scales through which human society is organized. Smith, for example talks about how local governments can bend scale to seek out international or supra-national partners for their local initiatives.

The units at different scales typically invoked in this human geographic research program, ranging from the body and household to the globe, are much like those of the other approaches summarized here. They form a hierarchy of territorial units of greater and greater geographic extent, with smaller scale units embedded within, and often nested inside, larger scale units. Yet the conceptualization of scale is quite different. First, the units at any scale are not fixed, as their boundaries can change over time (as when the country called "Yugoslavia" split from one nation-state into many). Their coherence as spatial units, their relative importance compared to other units at the same scale, and the importance of scales relative to one another, can all shift. The challenge, then, is to account for these changes, using sociospatial theory to unravel the complex and shifting relationship between societal processes and the spatiotemporal formations that both precipitate out of and also shape such processes. Second, causality does not begin at the lowest scale and work upwards. Much social and cultural theory rejects the methodological individualism common in the natural sciences, and in rational choice theory in the social sciences (Barnes and Sheppard, 1992). Instead of believing that social structures can be explained by aggregating together the actions of autonomous individuals, it is argued that social collectivities and individuals are mutually constituted. This implies that causality can run in all kinds of directions across (and within) scales. Third, instead of simply seeking to explain the world from some objective external position from which a final truth can be gleaned, it is argued that the fungibility of scale makes possible an emancipatory research program, asking how it is possible to shape scaled social processes in ways that empower all members of society and improve human livelihood possibilities.

A joint language for scale?

As can be seen in the chapters of this volume, a major difficulty in discussing scale is the disparate language used by geographers and others. Recently, there have been attempts to provide structure to the many definitions of scale. Lam and Quattrochi (1997) provide one such framework (Figure 12.1). They identify three aspects of scale of interest to geographers: spatial, temporal, and spatiotemporal, and offer a classification of scale "types", including cartographic, geographic, operational, and measurement. Based on human geographic work on scale, we have added a fifth class to their figure: The construction of scale. Note that the meanings of scale depicted here themselves are dependent on the context. For instance, the concept of measurement varies significantly from discipline to discipline and even within geography itself.

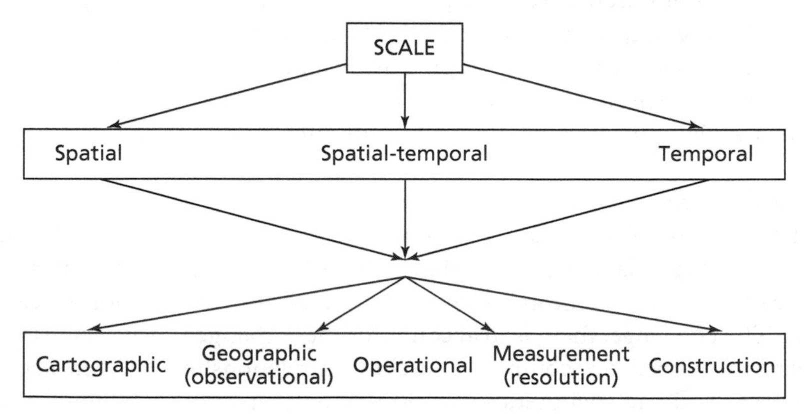

Figure 12.1 *The "meanings" of scale.*
Source: Drawn by authors, based on Lam and Quattrochi (1992).

Intersections

While it seems that geographic inquiry thus embraces very different approaches to scale, they are far from incommensurable. Indeed each can learn from other approaches, thereby strengthening its own thinking and creating a common ground of inquiry into scale. The conclusion that scale is not a fixed metric can improve scale-based analysis in geographic information science and physical geography. With some reflection, it is clear that scales and the units making them up are not fixed, but are an emergent property of complex dynamic natural and social systems. The extent and relative importance of watersheds at different scales, for example, are not determined once and for all, but are shaped both by societal processes (Swyngedouw, this volume) and by biophysical processes of landscape and climate formation. Even the fixed geometries of scale typifying the tradition of cartography are themselves dependent on societal processes that have influenced the geometries adopted to describe the Earth's surface. The choices of Cartesian coordinates and a square raster grid are by no means the most natural way to cover a spherical object like the earth in regular polygons, but reflect the strong influence of Euclidean geometry on cartographic norms. In addition, the nation-state has shaped the selection of topographic mapping scales, and the enumeration units (and categories) for census data. Regular polygons themselves conform with the data collection grids used in remote sensing and in digital imaging technologies, but are often a highly inadequate way to represent the irregular and fuzzily bounded scaled units of interest to physical and human geographers. One might be interested, for instance, in the appropriate scale to understand gang activity in Los Angeles, where boundaries are at best indeterminate. The idea that scales are constructed, and that we need to understand how this happens and how it changes, is thus relevant across all these traditions of geographic inquiry.

Perhaps the greatest difference between human and physical geography lies in the willingness of earth scientists to accept hierarchy theory as an adequate simplification of nature. Complex dynamic systems, such as those through which the structures and dynamics of nature and society, and their inter-relationships, emerge, do not naturally result in hierarchies of the kinds proposed by hierarchy theory. In addition, the explanation that hierarchy theory exists in nature because such systems are favored by processes of natural selection is at best speculative. Whether or not hierarchy theory is an appropriate way of simplifying the complexity of natural systems, it certainly is quite problematic as a way of describing social systems. While there may be some plausibility to the argument that interaction between different phenomena falls as physical distance increases in nature, where physical space poses a real barrier to animal movements, water flows, solifluction, and air mass gradients, the same is not true in human society. Thus even though social systems often exhibit scaled hierarchical characteristics, it does not follow that fast processes operate on small spatial scales and slow processes on large spatial scales (Goodchild and Quattrochi, 1997). Everyday experience attests to this; one of the fastest forms of interaction is via the Internet, where short-distance interactions are by no means the norm. If social space and distance do not conform closely to geographic space (Sheppard, 2002), then the correspondence between scale in time and space hypothesized in hierarchy theory (see Figure Intro.2b in this volume) is not appropriate for social phenomena. This must also raise questions, of course, about the appropriateness of hierarchy theory as a way of simplifying the complexity of nature, once we recognize that humans are a part of, and are not separate from, nature. This implies that the inapplicability of hierarchy theory to human society makes it also more difficult to apply to nature. Everyday experience of localized human activities that have immediate global consequences (the Chernobyl nuclear power plant explosion, fires in the Amazon, industrial pollution), testifies to this difficulty.

At the same time, approaches to scale in human geography can benefit from paying closer attention to work in geographic information science, spatial analysis, and physical geography. The poverty of available data (particularly on spatial flows, see Taylor, this volume), the importance of taking a historical perspective, the difficulty of simplifying complex social systems through the assumptions of hierarchy theory, and the current importance of qualitative research in the social sciences, all make it more difficult to engage in the kind of analytical research that currently dominates physical geography. Yet these difficulties should not become an excuse to avoid undertaking such research when it is beneficial. Databases are continually improving. Bayesian and exploratory statistics, as well as the advent of complexity theory in mathematics, mean that quantitative analysis need not be reductionist or positivist (Sheppard, 2001), but can contribute to and solidify the research program of critical human geography.

While the constructionist literature focuses on the contingent nature of scale, there are periods of time when scales do not change. As Smith (this volume) argues, such changes are periodic rather than continual. When analyzing scales that are not changing, human geographers often ask questions that are similar to those posed by physical geographers. For example, when discussing the shift from the dominance of the nation-state scale to glocalization, whereby supra- and subnational scales are now increasing in importance relative to the scale of nation-states, human geographers are engaged in identifying the pertinent scales, and in tracing the (changing) characteristic scales at which capital accumulation is organized. As Brenner (2001) argues, multiscalar analysis – examining the changing interrelations between scales – is a defining characteristic of constructionist approaches to scale. The insights gained from examining how relationships vary from one scale to the next are an important component of constuctionist approaches. For example, the observation that the relationship between risk of exposure to toxic chemicals and race is strong at the metropolitan scale but much less so at submetropolitan scales reveals the importance of suburbanization in shaping the relationship between race and pollution (see McMaster, Leitner, and Sheppard, 1997; Pulido, 2000).

Beyond Scale

Making sense of geographic scale is, thus, an important challenge for the human and earth sciences. It is vital to pay attention to what scale means, how to map it onto the earth's surface, and how scales are constituted, and change. The scaled nature of human and natural processes also poses challenges for explanation and analysis. A multiscalar theoretical and methodological strategy is vital for making sense of the world because different processes are dominant at different scales, and new insights into processes can be gained by examining them across a range of scales. It is for all of these reasons that geographic scale has become an important focus of research in and beyond the discipline of geography (White and Engelin, 1997; Jessop, 1999; Kates, Clark, Corell et al., 2001).

Yet, as Taylor and Leitner (this volume) both suggest, a focus on scale can pose dangers if it leads us to neglect other aspects of the spatial organization of nature and society, or to an attempt to collapse all such aspects into a scale-based analysis (see Brenner, 2001). Taylor discusses the dangers of treating one scale, the nation-state, as the pre-eminent scale for social science analysis (as in international relations, macroeconomics, and comparative sociology), and also shows how an understanding the nature of such subnational units as world cities needs to pay attention to interdependencies between those units. As Leitner argues, networks connecting cities across space, although themselves scaled, cannot be adequately captured by a scale-based analysis (in her case, one of the politics of scale). Such direct connections, reflecting and shaping the different situation or positionality of spatial units of a particular scale within larger scale geographic structures (Sheppard, 2002), have a distinct spatiality and reflect geographic processes that coevolve with, but cannot be reduced to, scale-based processes. In short, future research on scale and geographic inquiry should seek to refine not only our understanding of the nature and significance of geographic scale, but also the relationship between scale and other elements of a geographic way of knowing.

References

Barnes, T. and E. Sheppard 1992: Is there a place for the rational actor? A geographical critique of the rational choice paradigm. *Economic Geography*, 68: 1–21.

Brenner, N. 2001: The limits to scale? Methodological reflections on scalar structuration. *Progress in Human Geography*, 25(4): 591–614.

Church, M. 1996: Space, Time and the Mountain – How do we order what we see? In B. L. Rhoads and C. F. Thorn (eds), *The Scientific Nature of Geomorphology: Proceedings of the 27th Binghamton Symposium in Geomorphology held 27–29 September 1996*, London: John Wiley & Sons, 147–70.

Goodchild, M. 2000: Towards a location theory of distributed computing and commerce. In T. Leinbach and S. D. Brunn (eds), *Worlds of E-Commerce: Economic, Geographical and Social Dimensions*, New York: John Wiley & Sons, 67–86.

Goodchild, M. and D. Quattrochi 1997: Introduction: Scale, multiscaling, remote sensing, and GIS. In D. Quattrochi and M. Goodchild (eds), *Scale in Remote Sensing and GIS*, New York: Lewis Publishers, 1–11.

Jessop, B. 1999: Reflections on Globalization and its (Il)logics. In P. Dicken, P. Kelly, K. Olds and H. W.-C. Yeung (eds), *Globalization and the Asia Pacific: Contested Territories*, London: Routledge, 19–38.

Kates, R. W., W. C. Clark, R. Corell, et al. 2001: Sustainability science. *Science*, 292(5517): 641–2.

Lam, N. S.-N. and D. A. Quattrochi 1992: On the issues of scale, resolution, and fractal analysis in the mapping sciences. *The Professional Geographer*, 44(1): 89–99.

Levins, R. and R. Lewontin 1985: *The Dialectial Biologist*. Cambridge, MA: Harvard University Press.

McMaster, R. B., H. Leitner and E. Sheppard 1997: GIS-based environmental equity and risk assessment: Methodological problems and prospects. *Cartography and Geographic Information Systems*, 24(3): 172–89.

Monmonier, M. 1991: *How to Lie with Maps*. Chicago, IL: University of Chicago Press.

Pulido, L. 2000: Rethinking environmental racism: White privilege and urban development in Southern California. *Annals of the Association of American Geographers*, 90(1): 12–40.

Sheppard, E. 2001: Quantitative Geography: Representations, practices, and possibilities. *Environment and Planning D: Society and space*, 19: 535–54.

Sheppard, E. 2002: The spaces and times of globalization: Place, scale, networks, and positionality. *Economic Geography*, 78(3): 307–30.

Simon, H. A. 1962: The architecture of complexity. *Proceedings of the American Philosophical Society*, 106: 467–82.

Smolin, L. 1997: *The Life of the Cosmos*. Oxford: Oxford University Press.

White, R. and G. Engelin 1997: Multiscale spatial modeling of self-organizing urban systems. In F. Schweitzer (ed.) *Self-organization of Complex Structures: From Individual to Collective Dynamics*, Amsterdam: Gordon and Breach, 519–35.

Wood, D. 1992: *The Power of Maps*. New York: Guilford Press.

Wu, J. 1999: Hierachy and scaling: Extrapolation information along a scaling ladder. *Canadian Journal of Remote Sensing*, 25(4): 367–80.

Index

Page locators in *italics* denote tables and figures.